# Technological Progress in Latin America: The Prospects for Overcoming Dependency

**Other Titles in This Series**

*Brazil: Foreign Relations of a Future World Power,* Ronald Schneider

*Mexico's Economy: A Political Analysis with Forecasts to 1990,* Robert E. Looney

*The Future of Brazil,* edited by William Overholt

*The Politics of Chile: A Socio-Geographic Assessment,* César Caviedes

# Westview Special Studies on Latin America and the Caribbean

*Technological Progress in Latin America: The Prospects for Overcoming Dependency*
edited by James H. Street and Dilmus D. James

In light of the energy crisis and the problems of feeding a rapidly growing and urbanizing population, Latin America faces a critical need to acquire useful scientific and technological knowledge. Some writers have held that the region is hopelessly dependent on foreign governments and multinational corporations for its technology, and is thus doomed to fall behind. The contributors to this book, however, argue that there are possibilities for internal technological possibilities for internal technological development that have already been demonstrated in concrete instances and that can be expanded to alleviate the dependent status of the region.

The introductory section reviews the nature of technological dependency and makes a case for the stimulation of indigenous research and development; domestic obstacles as well as the potential for improvement are examined. The major portion of the book is devoted to case studies of successful innovations. The contributors, chiefly economists, have all had extensive field experience in Latin America and use empirical techniques to reach their conclusions.

James H. Street is professor and chairman of the Department of Economics at University College, Rutgers University. He has been a visiting lecturer at numerous Central and South American universities, and has conducted a wide range of field studies in Latin America.

Dilmus D. James is professor of economics at the University of Texas at El Paso. He also has served as a visiting professor in Mexico, and is the author of numerous journal articles on science and technology policy in developing areas.

# Technological Progress in Latin America: The Prospects for Overcoming Dependency

edited by James H. Street
and Dilmus D. James



Westview Press / Boulder, Colorado

*Westview Special Studies on Latin America and the Caribbean*



Published in 1979 in the United States of America by
Westview Press, Inc.
5500 Central Avenue
Boulder, Colorado 80301
Frederick A. Praeger, Publisher

Library of Congress Cataloging in Publication Data
Main entry under title:
Technological progress in Latin America.
**(Westview special studies on Latin America and the Caribbean)**
1. Technological innovations—Latin America—Addresses, essays, lectures. 2. Technology—Latin America—Addresses, essays, lectures. 3. Latin America—Economic conditions—1945- —Addresses, essays, lectures. I. Street, James Harry, 1915- II. James, Dilmus D.
HC130.T4T43 301.24'3'098 78-23980
ISBN 0-89158-255-X

Printed and bound in the United States of America

# Contents

# Figures

# Tables

*Chapter 11*

# Acknowledgments

Several of the chapters in this volume were prepared with the assistance of research grants, which the authors wish to acknowledge with gratitude.

Research for Chapter 2 on "The Extent of Technological Dependence in Latin America," by C. Richard Bath and Dilmus D. James, was supported by the University Research Institute of The University of Texas at El Paso.

Research for Chapter 4 on "Overcoming Technological Dependence in Latin America" and for Chapter 12 on "Technological Fusion and Cultural Interdependence," by James H. Street, was supported by grants from the Research Council of Rutgers—The State University of New Jersey.

Research for Chapter 8 on "Performance and Technology of U.S. and National Firms in Mexico," by Loretta Good Fairchild, was sponsored by the Program on Policy for Science and Technology in Developing Nations at Cornell University.

Research for Chapter 9 on "Organizational Structure and Innovativeness in the Pulp and Paper Industry of Mexico," by Viviane B. de Márquez, was supported by grants from the Social Science Research Council and the Consejo Nacional de Ciencia y Tecnología of Mexico.

Research for Chapter 10 on "Limited Search and the Technology Choices of Multinational Firms in Brazil," and Chapter 11 on "Adaptation by Foreign Firms to Labor Abundance in Brazil," by Samuel A. Morley and Gordon M. Smith, was financed by the National Science Foundation under grant No. GS-34581.

Gratitude is also expressed to the editors and publishers of the respective journals for permission to reprint material under copyright which appears as chapters in this volume: (1) Loretta Good Fairchild, "Performance and Technology of United States and National Firms in Mexico," *Journal of Development Studies* 14 (October 1977):14–34

(Chapter 8); (2) Samuel A. Morley and Gordon W. Smith, "Limited Search and the Technology Choices of Multinational Firms in Brazil," *Quarterly Journal of Economics* 91 (May 1977):263-87 (Chapter 10); (3) Samuel A. Morley and Gordon W. Smith, "The Choice of Technology: Multinational Firms in Brazil," *Economic Development and Cultural Change* 25 (January 1977):239-64 (Chapter 11); and (4) James H. Street, "The Ayres-Kuznets Framework and Argentine Dependency," *Journal of Economic Issues* 8 (December 1974):707-28 (Chapter 12). In addition, Chapter 7 is based on the author's *Organization for Rural Development: Risk Taking and Appropriate Technology* (New York, Praeger Publishers, 1978).

# Abbreviations

| | |
|---|---|
| AAAS | American Association for the Advancement of Science |
| AID | United States Agency for International Development |
| CAPES | Coordenação do Aperfeiçoamento de Pessoal de Nivel Superior (Coordination of the Formation of Higher Level Personnel—Brazil) |
| CDI | Comissão para o Desenvolvimento Industrial (Industrial Development Commission—Brazil) |
| CIA | United States Central Intelligence Agency |
| CNPq | Conselho Nacional de Pesquisas (National Research Council—Brazil) |
| COLCIENCIAS | Fondo Colombiano de Investigaciones Científicas (Colombian Fund for Scientific Research—Colombia) |
| CONACYT | Consejo Nacional de Ciencia y Tecnología de México (National Council of Science and Technology—Mexico) |
| CONICIT | Consejo Nacional de Investigaciones Científicas y Tecnológicas (National Council for Scientific and Technological Research—Costa Rica) |
| DEICOM | Departamento de Estatística Industrial e Comercial (Department of Industrial and Commercial Statistics—Brazil) |
| ECLA | United Nations Economic Commission for Latin America |

| | |
|---|---|
| FLACSO | Facultad Latinoamericana de Ciencias Sociales (Latin American Faculty of Social Sciences) |
| GDP | Gross domestic product |
| GNP | Gross national product |
| IBEC | International Basic Economy Corporation |
| IBGE | Instituto Brasileiro de Geografia e Estatística (Brazilian Institute of Geography and Statistics) |
| ICT | Instituto de Crédito Territorial (Territorial Credit Institute—Colombia) |
| IDB | Inter-American Development Bank |
| IDRIC | International Government Research Centre—Canada |
| INIC | Instituto Nacional de Investigación Científica (National Institute for Scientific Research—Mexico) |
| INPES | Instituto Nacional de Planejamento Econômico e Social (National Institute for Economic and Social Planning—Brazil) |
| IPEA | Instituto de Pesquisas Econômicas Applicadas (Institute of Applied Economic Research—Brazil) |
| ISB | Industrialized systems building |
| ISI | Import substitution industrialization |
| IVIC | Instituto Venezolano de Investigaciones Científicas (Venezuelan Institute for Scientific Research—Venezuela) |
| JV | Joint venture firms |
| LAFTA | Latin American Free Trade Association |
| LDCs | Less developed countries |
| LINCOTT | Liaison, interface, coupling, and technology transfer |
| MDCs | More developed countries |
| MNCs | Multinational corporations |
| NAS | National Academy of Sciences |

| | |
|---|---|
| NASA | National Aeronautics and Space Administration |
| NBER | National Bureau of Economic Research |
| OEA | Organización de Estados Americanos |
| OAS | Organization of American States |
| OECD | Organization for Economic Cooperation and Development |
| PREVI | Projecto Experimental de Vivienda (Experimental Housing Project—Peru) |
| PSR | Price sensitive range |
| R&D | Research and development |
| SBPC | Sociedade Brasileira para o Progresso da Ciência (Brazilian Society for the Progress of Science—Brazil) |
| SENA | Servicio Nacional de Aprendizaje (National Apprenticeship Service—Colombia) |
| SENAI | Servico Nacional de Aprendizagem Industrial (National Industrial Apprenticeship Service—Brazil) |
| TA | Technical assistance |
| UNAM | Universidad Nacional Autónoma de México (National Autonomous University of Mexico) |
| UNCTAD | United Nations Conference on Trade and Development |
| UNESCO | United Nations Educational, Scientific, and Cultural Organization |
| UNIDO | United Nations Industrial Development Organization |
| USAID | United States Agency for International Development |
| VITA | Volunteers in International Technical Assistance |

# About the Contributors

**C. Richard Bath** is associate professor of Political Science at the University of Texas, El Paso. He has published articles on dependency in Latin America, the U.S.-Mexico border, and environmental issues. He is currently researching environmental and energy policy in Latin America.

**Viviane B. De Marquez** is professor of sociology at El Colegio de Mexico and a participant in that institution's Science and Technology for Development research program. She is currently researching technological innovation in the Mexican food industry.

**Lorretta G. Fairchild** is assistant professor of economics at Nebraska Wesleyan University and consultant for the Program on Policies for Science and Technology in Developing Nations at Cornell University. Her current research is focused on technology transfer in Brazil, Colombia, and Mexico.

**David Felix** is professor of economics at Washington University in Saint Louis. He has served as consultant with economic research institutes in Santiago, Buenos Aires, São Paulo, and Quito, and has been visiting fellow at Harvard and the University of Sussex. His current research includes projects in technology and economic growth, consumer dynamics and economic growth, and Mexican income trends.

**Allen D. Jedlicka** is associate professor of organizational behavior at the University of Northern Iowa. He is the author of a book on organizational techniques for technological adaptation in rural development. His current research includes technology transfer, appropriate technology, and rural development in Latin America.

**Samuel A. Morley** is professor of economics at Vanderbilt University. He has written extensively on the economic growth and development of the Brazilian economy.

**Gordon W. Smith** is professor of economics at Rice University. He has served as consultant to the Brazilian Government, the United States State Department, and the U.S. Treasury Department. In addition to his work on choice of technology, he has published a book and several articles on the international exchange of raw commodities.

**Arthur H. Rubenstein** is professor of industrial engineering and management science at Northwestern University. He has established the Program of Research on the Management of Research and Development at Northwestern and has conducted and directed research on various aspects of technology transfer.

**W. Paul Strassmann** is professor of economics at Michigan State University. He has served as consultant to the Agency for International Development, Brookings Institution, the National Academy of Science and Engineering, the United Nations and the World Bank. He has also been a visiting scholar at the London School of Economics and the International Labor Office in Geneva. In addition, Strassmann is the author of the recently published book, *Housing and Building Technology in Developing Countries.*

# 1
# Introduction

*James H. Street*
*Dilmus D. James*

Recent studies of the increasingly critical developmental problems of Latin America have stressed the dependent relationships between the region and industrially advanced countries. In emphasizing the dominance of powerful foreign governments and multinational corporations in Latin American economic and political affairs, technology, and cultural evolution, the Dependency School has tended to minimize the role of domestic factors, including the distribution of internal resources, the process of historical conditioning, and the capacity of Latin Americans to exercise initiative in their own behalf.

This volume, although it did not arise in the heat of controversy, attempts to redress this imbalance. Without ignoring the often severe impact of outside forces on Latin America, the writers of these essays are deeply concerned with the internal forces—particularly those utilizing more advanced techniques—that have shaped the development process and will condition it in the future. They perceive that the application of new technologies to the more effective use of resources will provide the principal hope for Latin Americans' meeting the problems of their future. Some are less optimistic than others, but together they reject the deeply pessimistic view that Latin America is hopelessly dependent on decisions made exclusively in the world's power centers.

The contributions to this volume were selected from among a large number of papers presented at recent meetings of the International Studies Association, the Latin American Studies Association, the Southwestern Social Science Association, and the Rocky Mountain Council for Latin American Studies. The collection is representative of the work of an informal network of research collaborators who have addressed themselves to questions concerning the historical lag in technological development in Latin America, the increasing dependence of the region on outside sources for useful knowledge, the diffusion of new techniques and processes within the region, and the gradual emergence of genuinely domestic research and development. Most of the investigators are economists by training, but they have not hesitated to draw on other disciplines for findings relevant to their studies of the complex process of technological evolution. Each contributor has had extensive field experience in Latin America.

Even though these writers occasionally disagree on the strategy and tactics of progress, their outlook is constructive. They believe that Latin American nations can and should take measures to acquire scientific and technical knowledge on more favorable terms, to foster their own research and development for the solution of concrete problems, and to thereby achieve a better ecological relationship among domestic resource endowments, markets of varying size, social institutions, and human needs.

The essays are divided into two groups: those relating primaily to the general nature of technological dependency and development in Latin America and those describing specific cases of the generation and diffusion of new techniques within countries or sectors of the region. There is considerable diversity in scope and approach, but a pervasive underlying theme focuses on the rich opportunities for accelerating the growth process through the creation and adoption of appropriate technologies developed within the world scientific community.

It is useful to contrast this approach with that of the Dependency School. In general, the writers of this school regard the development of industrialized countries (the "center" or "metropole") and the deteriorating condition of underdevelopment of the peripheral countries as inextricably linked. The retrograde process of underdevelopment, they believe, cannot be understood except as an aspect of the growth of the industrially advanced world, which feeds on and exploits the resources and incomes of the less developed world.

Secondly, they believe the basic decisions about the pattern of organization and economic policies within the peripheral countries are heavily influenced by direct intervention by the center or by imitation among the controlling classes of the less developed countries themselves. Local governments and business groups are regarded as having been coopted and are thus assumed to be acting as agents of foreign governments and multinational corporations. Even educational institutions and research centers are described as slavishly imitative rather than boldly innovative in their approaches.

Finally, the more radical dependency writers hold that the increasingly deleterious effects of these relationships on the less developed world are the result of a deliberate and concerted effort on the part of the industrial countries—an international exploitative conspiracy that holds the victims powerless to affect their own destinies.

The projections drawn from this analysis are generally bleak. They point to increasing misery of the world's poor or to violent overthrow of existing governments and economic institutions. Members of the Dependency School discount alternative strategies that depend on local initiative (other than sweeping revolution) because they regard all other possibilities as merely meliorative and tending to reinforce the basic exploitative relationship. This attitude may account for their failure to explore the possibilities for genuinely indigenous science and technology and the ways in which such activity may be fostered and applied to the development process.

The writers in this volume take a more hopeful view. Although aware that there is a massive technological gap between Latin America and the advanced industrial countries, they nevertheless reject the notion that this condition is doomed to worsen and that substantial improvement in the condition of the working classes in Latin America is beyond possibility. The resources of the region are far from fully exploited, and the fruits of technological progress are available to others besides those who have had early access to them.

Secondly, these writers believe that within the diversity of situations in Latin America people can exercise their own volition and find applications of new methods not subject to control and manipulation from abroad. Dependence is a matter of degree, and some countries, such as Brazil, have already shown considerable independence in directing their internal development. There is more than one road to progress; the history of a variety of economically diversified countries illustrates that an accumulation of modest improvements in many directions may have broadly progressive consequences.

Thirdly, these writers do not accept the concept of a concerted conspiracy on the part of the peoples of the advanced world to hold back the underdeveloped world. There exists an international community of researchers and innovators whose interest in science and technology transcends national boundaries and interests. Useful knowledge is generated by public agencies, foundations, and educational institutions that promote its worldwide dissemination rather than seek to hold it under control for maximum profit. And there are competing sources of privately generated research in many countries that the less developed countries may draw on; often, they may use the resources of international lending agencies to finance the utilization of such technology. In sum, science and technology tend to become universally available, and they offer an escape from hopeless dependency and incessant foreign exploitation.

In the first group of essays dealing with the general nature of technological dependency in Latin America, C. Richard Bath and Dilmus D. James consider how extensive the technological gap between the region and the industrialized world is, and whether it can be reduced. After reviewing the principal complaints about the conditions under which technology is acquired, the authors attempt to demonstrate that the scientific and technological gap is not as debilitating as it first appears. They point out that there are a variety of internal policy measures available to decision makers in each country that would encourage the adoption, adaptation, and development of more appropriate technologies. The fact that some adaptation takes place even under the most discouraging conditions suggests that improvements in competitive conditions and factor markets would significantly expand the creative process of selection, alteration, and development of new techniques.

David Felix treats the diffusion of technology in Latin America as an offshoot of the Industrial Revolution in Europe and the United States, with its profound effects on international trade and investment. To explain the dual economies of Latin America and the limited diffusion of advanced technology in the region, he asserts that it is necessary to look for something other than the free working of international markets. He divides the history of technological development in the industrial countries into two periods: from the 1860s to the 1920s, and the recent period thereafter. He concludes that in each historical phase, distinguished by different methods of generating new discoveries, Latin America was unable to participate in time to develop its own sources of innovation. The reasons had little to do with conditions imposed by a dependency relationship, but rather resulted from the peculiar evolution of local institutions and modes of production (which Felix describes in detail).

James H. Street, who considers the means of overcoming technological dependence, also addresses internal factors in Latin America that have impeded the cultivation of attitudes and practices necessary to the stimulation of invention and discovery. He looks particularly at educational institutions, which at times responded to the example of useful innovations from abroad only to lose their technological orientation as political influences intervened. He suggests that there are positive influences in the Latin American culture that could be utilized to surmount the continued need to depend on borrowed ideas and techniques.

Dilmus James concludes this part of the volume with an examination of the economic case for more indigenous research and development in less developed countries. He asserts that many techniques needed for particular local purposes are not available in the backlog of knowledge of the developed world and therefore, may best be originated or adapted in the region. He cites instances in which the costs of transferring technology are underestimated and others in which the returns to indigenous research and development are also not fully appreciated. Finally, he argues that there are extraeconomic reasons for supporting more intensive research and development effort within the less developed world which will ultimately have a cumulative effect.

Part 2 of the volume presents case studies in internal technological diffusion, drawing on research conducted in several economic sectors and countries in Latin America. Allen D. Jedlicka and Albert H. Rubenstein report on a survey conducted in Colombia to discover what factors constitute the principal barriers to acquisition of technological information and adoption of new techniques by Colombian industrialists. The chief difficulty perceived by these businessmen is the poor training of extension agents in areas in which they need information. This often leads them to seek information from abroad. Some respondents also expressed fear that the government, in order to improve the scope of domestic information services, would utilize the return flow of information about industry to increase controls and impose higher taxes. The authors conclude that there may be more effective means of increasing the speed and scope of information acquisition by Colombian industry.

Turning to the agricultural sector, Allen D. Jedlicka reports on a successful program for the transfer of information to peasant farmers in the state of Puebla, Mexico. He concludes that the success of the program rested upon three factors: (1) the creation of a feeling of trust between the disseminators of technical information and the peasant recipients; (2) the participation by local farmers in the decisions affecting their lives; and (3) the availability of effective means of mass communication at low cost. He describes the usefulness of Rensis Likert's "linking-pin" model in this experiment.

In an empirical study bridging five years, Loretta Good Fairchild found that the characteristics of industrial firms in Monterrey, Mexico, differed significantly from those described in the "dependency" literature. Not only are firms under Mexican management in a broad range of industries performing comparably with foreign subsidiaries in terms of profitability, growth, and exports, but they are at least as innovative in introducing new products and productive processes. Moreover, they appear to be relying substantially on domestic institutions, and particularly on resources internal to the firm, to generate new technology. Fairchild's findings suggest that a new national industrial base, not subordinated by foreign competition and control, is becoming established in Mexico.

After a brief review of the literature on the transfer of technology by multinational firms operating in Latin America, the contribution of Viviane B. de Márquez summarizes her research findings on the incidence of innovations in the Mexican pulp and paper industry. Her investigation, which entailed personal interviews with management, plant supervisors, and equipment operators, does not support the contention that transnational firms tend to be less innovative than their national counterparts. She finds that multinational pulp and paper firms introduce new product lines more frequently, exploit a wider range of information sources, and are generally more innovative. The author is careful to qualify her conclusions, however, because of the impossibility of separating size of firm as a controlling factor. (Larger firms tend to be more innovative and the transnational firms are among the largest in the sample studied.)

In a pair of related studies conducted in Brazil, Samuel A. Morley and Gordon W. Smith investigated how multinational firms make choices concerning the production techniques they will use. In a survey of production managers in thirty-five Brazilian firms, they found that low wage costs play an insignificant part in influencing the technological choices of multinational firms. Rather, managers make discretionary judgments on labor-saving techniques they are already familiar with in their home countries. Economies of scale far outweigh low labor costs as the determinant of machine choice and labor use.

Using regression analysis, Morley and Smith found significant differences among technologies employed in Brazil by foreign firms of different nationalities and between technologies used by foreign firms and those employed by Brazilian firms. These differences are explained by the accumulated experience of the respective managers. In the permissive Brazilian environment, with rapid growth, formidable import barriers, and little price competition, managers of foreign multinationals can readily meet their profit targets without having to search for less familiar labor-intensive techniques. For Brazilian firms, on the other hand, their more limited experience necessitates an extensive and sometimes costly search for more labor-intensive alternatives.

In their companion study, which is based on a close investigation of the differences between metalworking firms in the United States and their subsidiaries in Brazil, Morley and Smith raise questions about whether or not multinational firms have adapted their production techniques to employ more labor and less capital in Brazil, and if so, for what reasons. They also ask whether the Brazilian government could have found ways to induce foreign firms to employ more labor. Confirming the results reported in Chapter 10, they conclude that the subsidiaries of U.S. metalworking firms do, indeed, employ different techniques in Brazil than their parent companies use at home, but that these differences do not stem from the availability of cheap labor. Rather, they are closely related to differences in the scale of operations required by the smaller Brazilian market. The policy implication is that the Brazilian government could do little to increase the employment of Brazilian workers by wage controls or subsidies. The authors suggest that the government could probably be more effective in stimulating employment if it encouraged the production of commodities requiring more labor-intensive methods, rather than trying to induce companies to use more labor to make products they are already geared up to produce efficiently by machine.

In contrast with the dynamic growth of Brazil during the past two decades, Argentina has contended with economic stagnation. In an attempt to explain the slowdown of the Argentine economy, James H. Street has centered on the forces affecting technological development within that country. He finds that Argentina's early dynamic growth period can be explained primarily by a process of "technological fusion" in which foreign influences played a large role. After World War I, institutional rigidities set in, and Argentina failed to domesticate and propagate the technological process that had been so powerful in promoting early progress. As a consequence, the Argentine economy failed to "take off" into self-sustaining growth and has been subject to stop-go cycles in recent decades. Street believes that Argentina will continue to depend on foreign sources for much of its new technology, but that the country has the capacity to rebuild its educational and research institutions; therefore, it could become part of the *interdependent* world network of scientific and technological exchange.

In the final chapter, W. Paul Strassmann looks at the critical housing sector of the Latin American economies. He reviews the use of prefabrication, or industrialized systems building, to relieve the housing shortages in Puerto Rico and Colombia. Prefabrication has had only limited success in Puerto Rico and has failed completely in Colombia, where the quality of houses was too low, the volume of construction too small, and the costs of building too high. He considers a variety of other innovations in construction materials and methods that are more closely related to traditional building techniques. In general, these innovations have the advantages that they lower costs, expand the use of local labor, and rely more heavily on locally available materials than the prefabricated methods using high automation. Strassmann believes that the goals of providing more and better housing on the one hand, and more employment on the other, are compatible in Latin America, and that the search for appropriate technologies to meet these goals should take precedence over the imitation of automated methods which are better adapted to the industrially advanced countries.

The studies contained in Part 2 of this volume are indicative of a wide range of research now going on that concerns itself with the more effective use of technology and the genuine domestication of the entire technological process in Latin America, as well as in other parts of the underdeveloped world. Before we conclude that these regions are condemned to dependence and exploitation by the more advanced countries of the industrial world, we might do well to look at the history of international technological diffusion during the past two centuries as an indication that the process is still going on and may yet have profound results.

# Part 1
# The General Nature of Technological Dependency

# 2
# The Extent of Technological Dependence in Latin America

*C. Richard Bath*
*Dilmus D. James*

The matter of acquiring technology has become a crucial issue to those concerned with the political economy of development. The International Labor Organization has sponsored a series of studies on technology transfer in conjunction with its World Employment Program; the United Nations Conference on Trade and Development has created a section specializing in the topic; and the United Nations Institute for Training and Research has produced a series of research reports, each dealing with a specific aspect of technology transfer.

With specific relation to Latin America, the Organization of American States has had a regional program on scientific and technological development for some years and has initiated a pilot project for the study of the transfer of technology. *Comercio Exterior,* a monthly publication of the Banco de México de Comercio Exterior, has become perhaps the leading outlet for scholars writing on science and technology policy for developing nations. The United Nations Economic Commission for Latin America set up an Expert Committee on the Application of Science and Technology for Development late in 1974.

In addition, a series of meetings in Latin America have dealt with technology policy; two of the more prominent were the 1965 Conference on the Application of Science and Technology to the Development of Latin America in Santiago, sponsored by UNESCO and the Economic Commission for Latin America, and a similarly sponsored conference in Brasilia in 1972. Transfer of technology was a major issue discussed at a 1974 conference held by twenty-five foreign ministers from the western hemisphere in Mexico City, and a statement regarding such transfers was included in the resulting Declaration of Tlatelolco.[1]

The growing literature on science and technology policy in Latin America has generated considerable controversy. After reviewing the major areas of contention surrounding technology transfer, we believe that one important dimension of achieving a viable overall Latin American science and technology policy involves taking actions over which Latin America has at least partial control. There is, we feel, a danger in becoming so preoccupied with external forces that meaningful internal measures are overlooked.

## The Width of the Gap

Technological dependence is often associated with a technological gap between less developed countries (LDCs) and more developed countries (MDCs). One encounters rough estimates such as that by Felipe Herrera: 75 percent of the economic growth in MDCs is due to technological advance and 25 percent to population increase, whereas these proportions are reversed for LDCs.[2] Others use gross national product (GNP) differentials as a reflection of a technology gap. Between 1950 and 1966 the difference between the GNP for Latin America and that for the United States and the European Common Market widened by 144 percent and 236 percent, respectively.[3] One observer put Latin America's investment in research below one-tenth of one percent of GNP in the late 1960s and claimed that Latin America's percentage of students enrolled in scientific and technological fields is the lowest in the world.[4] Another estimate for the same period shows research and development (R&D) expenditures as a percent of GNP: 0.3 percent (Argentina); 0.24 percent (Bolivia); 0.18 percent (Brazil); 0.26 percent (Colombia); 1.3 percent (Peru); and 0.16 percent (Venezuela).[5] UNESCO data for the late 1960s and early 1970s show that, typically, Latin American countries spend from 0.1 to 0.4 percent of their GNP on R&D.[6] This compares to a range of 1.0 to 3.0 percent for most developed countries.[7] From a sample comparing five MDCs (the United States, the United Kingdom, West Germany, France, and Holland) with five LDCs (India, Malaysia, Ceylon, Chile, and Tanzania), it appears that twenty to fifty dollars are spent annually on agricultural research per farm family in MDCs, whereas the range in LCDs is between five cents and two dollars.[8] Occasionally, an imaginative comparison emerges, such as the number of geologists per million square kilometers.[9] Figures on individual industries can be revealing: in 1971, LCDs had an adverse trade balance of $674 million in pharmaceuticals, considered a highly research-intensive industry.[10] Imperfect and incomplete as they are, perhaps the most revealing figures deal with balance-of-payments items directly tied to technology transfer. In the mid-1960s, LCDs received less than 1 percent of this total and paid 8 percent.[11] For the late 1960s, Latin America's net deficit on technology payments was estimated at over $300 million.[12]

## A Critique of Conventional Theory

What difference does all of this make to science and technology strategy in Latin America? Why cannot the accumulation of scientific and technical know-how be regarded as an investment, and international transfers of technology be handled adequately by the classical law of comparative advantage? Acquirers of technology in relatively technology-scarce economies may compare the costs of producing it locally or importing it from abroad. If the latter is the least-cost option, and the prospect of future gains from employing the new technology is sufficiently enticing, the purchaser acquires the technology from abroad by buying references, hiring technical experts, sending students abroad to study, contracting for a license to produce under a patent agreement, or using some other method or combination of methods. Presumably, in a neoclassical competitive world with flexibilities, no external economic effects, no troublesome time lags in adjustments, and with tendencies toward equilibria, technology embodied in capital equipment or in the form of books, blueprints, films, and human brains would flow rapidly to capital-scarce areas, thus benefiting all concerned in a way reminiscent of Portugal's swapping wine for English cloth in Ricardo's celebrated example from the early nineteenth century.

Before identifying the major objections to this neoclassical model, we would like to note some minor criticisms. First, if Latin Americans regard a pronounced technological gap as a sign of inferiority, it is perfectly logical for them to import less technology and produce more of it domestically, even if production costs are higher at home. The higher costs may be ascribed to the desire to avoid appearing inferior. Domestic science and technology are then considered partly items of consumption.

In addition, knowledge is subject to an imperfection in the marketplace that is difficult to remedy, since one cannot know exactly what he is purchasing until the transaction is completed. This is what Kenneth Arrow called "the fundamental paradox" of knowledge[13] or what Charles Cooper, using a more colorful expression, dubbed "the pig in a poke" thesis.[14] This imperfection will have ramifications, as we shall see when we discuss monopolistic power in the transfer of technology.

Finally, Raúl Prebisch has attacked the theory of comparative advantage on the ground that raw-commodity-producing and exporting countries are unable, for a variety of reasons, to capture all of the gains from cost-reducing innovations in traditional export products.[15] We will not elaborate on the Prebisch thesis, partly because his policy remedies involve institutional adjustments rather than modifications in science and technology policy, and partly because there is a body of literature that seriously questions Prebisch's conclusions.[16] We will consider market imperfections in the transfer of technology, which we regard as more significant. These can be grouped under (1) monopolistic practices, (2) inappropriate technology, and (3) discontinuities and disequilibria.

Much of the path-making work on monopolistic practices associated with technology transfer has been done by Constantine V. Vaitsos. In an intensive study of business practices of foreign firms and their local affiliates in Colombia he uncovered a variety of undesirable practices.[17] He found that purchasers of proprietary technology are at a substantial disadvantage in the bargaining process due to a lack of knowledge of alternative technologies and a restricted number of alternative sources with which to bargain. Prices for inputs acquired by foreign affiliates, often under tying agreements with the parent firm, are frequently much higher than quoted world prices. Licensees are often barred from exporting or must agree that any local innovations on a product belong to the parent company. The classical strategy of market discrimination is used to take advantage of differences in demand conditions and bargaining strengths among purchasing countries or, for that matter, individual firms in LDCs. Exorbitant charges to the affiliate (even to a wholly owned subsidiary) for technology transfers embodied in intermediate inputs, capital equipment, or managerial arrangements and engineering contracts are used to shift profits from subsidiary to parent company—to the detriment of the host country's balance of payments and tax revenues. Furthermore, Vaitsos views the patent system in its present form as an institutional arrangement curtailing the flow of useful technology to LDCs and driving up the price for this technology.[18]

There is general agreement that the technology transferred to Latin America and other developing areas is better able to accommodate the resource endowments of MDCs, where labor is relatively scarce and capital is relatively abundant. Such technology usually has a poor fit to labor-abundant, capital-starved Latin America. Furthermore, there is evidence that too little is done in LDCs to adapt transferred technology to different factor supplies or smaller market sizes. This, it appears, is particularly true of foreign firms. Grant L. Reuber found that in only nineteen of seventy-eight investment projects by multinational corporations (MNCs) sampled were technologies adapted to conditions in LDCs.[19]

In their study of adaptation by foreign firms to labor abundance in Brazil (Chapter 11), Morley and Smith conclude that among thirty-five foreign firms little was done to alter production techniques. Wayne Yeoman found similiar tendencies for thirteen MNCs in his study.[20] Another investigation of 1,484 affiliates of U.S.-based MNCs found little systematic difference in labor and capital intensity of techniques employed by MDC and LDC affiliates.[21] This does not imply that adaptations fail to take place at all; we will adduce some supporting evidence that they do. Yet we have no quarrel with the proposition that technology is often transferred in an essentially unmodified form, especially by foreign enterprises.

Several consequences follow from the transplantation of capital-intensive, large-scale technology; most of them are not favorable for Latin America. Since the latest vintage of technology requires an increased scale of output, when it is set down in smaller Latin American markets, the result is unused productive capacity. In the mid-1960s, a study of seventeen integrated steel mills in Latin America revealed an average of 48 percent unused capacity in the blooming mills.[22] Only four of the plants utilized more than 80 percent of capacity. During the same period, only 71 percent of installed capacity in paper and 68 percent of installed capacity in pulp were utilized.[23] Tibor Scitovsky cites figures ranging from 46 to 65 percent average industrial capacity utilization in Argentina, Chile, Colombia, and Ecuador.[24]

Secondly, capital-intensive production techniques distribute income more to the advantage of the owners of capital and the professional working class than to the advantage of the lower skill-range of the labor force. This further exacerbates, or at best does little to alleviate, the already pathological socioeconomic condition, in which the lowest 20 percent of the population receives an estimated 3.1 percent of income, while the top 5 percent receives 33.4 percent.[25] Furthermore, foreign firms that transfer technology tend to produce products that are, in the Latin American context, semiluxury or luxury goods that do little for the needs of the bulk of society. The importation of more sophisticated technology tends to raise the level of importation of inputs to support industrial growth. This puts additional pressure on balance-of-payments problems. Finally, the capital-intensive industrial expansion of Latin America has been very disappointing in terms of absorbing labor productively.

Some Latin American scholars, the most prominent of whom are Celso Furtado and Osvaldo Sunkel, picture these characteristics as a constellation of interrelated, self-sustaining forces and the prime mover as the MNC.[26] The scenario is as follows: foreign investment is attracted to Latin America, stimulated in part by domestic measures designed to encourage import-substituting industries. Due to some combination of lack of economic incentives, managerial inertia, and lack of knowledge about alternative techniques, technology is transferred with only moderate adaptation to local factor proportions and market sizes. Perhaps because of managerial inertia or because a foreign firm wishes to help pay for its product development costs, the products produced by foreign firms are typically consumer goods or capital equipment required to produce consumer goods, and these goods are introduced in a relatively unmodified form. Industrial growth, based on technology and products from affluent nations, tends to distribute income in favor of wealthier groups, a situation which further stimulates demand for luxury goods and attracts more foreign investment. Impressive rates of industrial expansion can be observed, yet, for the most part, society as a whole is not participating. The dynamic cycle tends to perpetuate and entrench itself as advantaged local groups become consciously or unconsciously supportive of the process. Given the marginalization of much of the population, any mechanisms for social protest or reform are too feeble or isolated to influence the industrial elite and hangers-on who champion this brand of economic growth.

**Barriers to Indigenous Research and Development**

An entirely different set of problems arises if we suppose there are lucrative long-run pay-offs to indigenous research and development (R&D). Yet there are distinct barriers to building up such capabilities within Latin America.

When there are large economies of scale in R&D activities, a short-run comparative advantage may perpetuate or even accentuate a technological gap indefinitely.[27] Something akin to the infant industry argument applies here. Not only conventional economies of scale but also gains in efficiency due to accumulated experience or "learning by doing" may bless early entrants into R&D with an initial cost advantage difficult to surmount in the short run. The same may occur if there are significant external economies to individual R&D projects. If a large number of projects is carried out simultaneously, for instance, a firm may be able to utilize new information that is not pertinent to a particular project but valuable to other R&D projects. Other forms of "cross fertilization" are well known in the scientific community.

Any one of these phenomena or some combination of them could freeze out Latin American R&D efforts for the foreseeable future, even though long-term gains might justify a more heroic expenditure on indigenous science and technology. Furthermore, leaving decreasing costs aside, some scientific endeavors require such enormous initial expenditures that all of Latin America combined could not afford the initial ante to enter the game. Examples are computer research, space exploration, supersonic aircraft, and mining the ocean depths.

A further type of barrier to indigenous R&D in Latin America mentioned frequently in the literature is the "critical minimum" or "critical mass" of scientists or technicians necessary to make R&D effort worthwhile. Conceptually, we may distinguish four minimum-effort situations. First, there may be a "micro-minimum," in which each individual research team with members closely interrelating on a research project must reach some sufficient size in order to be effective. Naturally this will vary with the nature of the research, and, for that matter, with the capabilities of the team members.[28] Second, a "meso-minimum" can be distinguished when research projects are interdisciplinary and demand a certain amount of interaction and input from several fields of research. For the pharmaceutical industry this has been estimated at about 200 "employees," presumably scientists and technicians.[29] Third, a "macro-minimum" describes a threshold beyond which the scientific and technological community can begin to have some concrete influence on the amount of resources devoted to research activities and supporting infrastructure. Mexico, for instance, appears to have arrived at this stage with the development of a comprehensive budgetary blueprint extending to the year 1982.[30] Finally, beyond a "mega-minimum," scientific and technological endeavors become well recognized by the society at large and affect social institutions, cultural values, and ways of viewing man's physical environment.

In summary, there are various discontinuities offering resistance to the development of Latin American technological capabilities. Some dynamic cost advantages that may accrue to latecomers, such as Japan, actually place the *later* latecomers at a decided disadvantage.

There are additional complaints levied against the present network for transferring technology, but those mentioned above appear to be the ones most objectionable to Latin Americans. The concept of "dependence" implies that there are determining forces external to Latin America. This we certainly do not deny. If there is to be a reordering of the international patent system, increased relevance and efficiency in the technology component of foreign aid, or some participation by Latin American nations in such activities as satellite communications research, then obviously external forces are important; they are indeed in a controlling position. Moreover, outside factors—such as the U.S. Central Intelligence Agency (CIA), Export-Import Bank, and international lending agencies—can intentionally or unintentionally influence the relative potency of internal forces in Latin

America. The MNC, especially, displays a particularly dazzling virtuosity in its methods for influencing internal decisions, as recent disclosures of corporate bribing of government officials attest.

We would now like to focus our analysis on the contention that, despite these powerful and pervasive forces, there are measures that Latin Americans can take that will encourage more rational selection of existing technology, more vigorous local modification of transferred technology, and active innovation within Latin America. Our argument centers on the contention that some movement toward a more competitive market with prices for factors of production closer to their real opportunity costs would render technology transfer and local technological activity more efficient.

## Cases of Domestic Innovation

Before developing this theme, however, we must point out that Latin America's situation may not be quite as hopeless with respect to acquiring technology as one would think from reviewing various indicators of the technology gap cited above.[31] Loretta Fairchild's study (appearing in this volume in Chapter 8) compares U.S.-affiliated firms and Mexican firms in Monterrey, Mexico and concludes that the Mexican entrepreneurs were not at a substantial disadvantage in gaining access to and acquiring technology. Interestingly enough, the Mexican entrepreneurs *felt* that they were disadvantaged, but the facts did not bear this out. While it is undoubtedly true that Monterrey is atypical because of its proximity to the United States, its traditional independence from political decisions in Mexico City, the lack of a tradition of landed elites around Monterrey, and an early start on industrialization resulting from a demand for Mexican products during the U.S. Civil War,[32] the study indicates that under favorable conditions Latin American entrepreneurs need not be completely frozen out of the game.

Recent legislation regulating the importation of technology into some Latin American countries is further evidence that internal actions can be taken. Enacted by the Andean Pact Group (in 1970), Brazil (in 1970 and 1971), Argentina (in 1971), and Mexico (in 1973), these laws permit the regulation of existing and future contracts entailing the importation of new technology.[33] Many provisions of the Mexican law have some resemblance to anti-trust legislation of the United States and some European countries. Clauses in contracts that prohibit exports or contain tying agreements, for example, are ordinarily excluded. Since special boards are able to decide on the applicability of the general rules to each particular case, it was feared in Mexico that policy would prove too flexible and even be favorable to foreign investors and sellers of technology.[34] It appears, however, that Mexican authorities have been rather strict. Based on the life of contracts involved, a Mexican study calculates that as of July 1976, 5.6 billion pesos were saved in payments for technology.[35]

The elimination of export restrictions has also had significant effects. Richard Morgenstern and Ronald E. Muller believe that contractual export restrictions, imposed as a condition for obtaining technology, may explain why the export behavior of Latin American firms during the late 1960s showed such low participation by the affiliates of MNCs.[36]

## The Importance of Factor Prices

Why is much of the technology imported into Latin America in a relatively unmodified form? Why is there not more local adaptation of imported technologies? Why is there so little development of new technologies within Latin America? Relative factor prices do not appear to bring about a more rational use of technology that would normally be expected in a market economy. A significant part of the problem may lie in market imperfections that do not provide the incentive to use more appropriate technologies. Some economists are convinced that on the one hand in poorer countries wages for lower skilled laborers are above the real-opportunity cost of productively employing them and, on the other hand, the productivity of capital is above the going rate of interest. A voluminous literature has grown up to measure and explain the extent of these distortions. The main effect is rather obvious: these imperfections in factor prices tend to skew the choice of technique toward a more capital-intensive method of production. The existing market realities, as viewed by the entrepreneur, may make the importation of essentially unmodified technology from affluent societies a rational choice, even though the social cost is high.

In addition, there is a range of policy measures in Latin America designed to encourage import-substitution. A protected environment allows firms the luxury of not bothering to expend managerial time, effort, and ingenuity in searching for alternatives to the newest vintage of technology, undertaking more extensive adaptation, or developing new technologies locally. According to Little, Scitovsky, and Scott, the effective rate of protection for manufactured goods was 162 percent for Argentina (in 1958), 118 percent for Brazil (in 1966), and 27 percent for Mexico (in 1960).[37] Bela Balassa and associates estimate effective protection rates for manufacturing, depending on the method used for estimation, to be 113–127 percent for Brazil (in 1966), 158–182 percent for Chile (in 1961), and 20–32 percent for Mexico (in 1960).[38] Given these circumstances, it is not worth the opportunity cost of managerial effort to look for ways to acquire technology with a better fit to factor endowments.

This raises a further set of questions, however. Suppose that institutional changes were made in tax policy, labor legislation, and financial arrangements that would cause wages and the rate of interest more nearly to reflect the social opportunity cost of labor and capital equipment. What assurance do we have that this would alter the technology selection, innovation, and R&D behavior of Latin American entrepreneurs, whether they are representatives of MNCs or local managers? Strictly speaking, we have *no* assurance. Yet there are indications that we could expect at least some shift to a more socially beneficial technology policy in an economic atmosphere that provides a greater incentive to do so.

First, without disregarding our earlier proposition that technology is generally transferred in an unmodified form, especially by MNCs, we stress here that this does not mean that there is *no* adaptation. However meagerly, some firms, including MNCs, *do* modify technology. In the previously cited study of Grant L. Reuber, fifty-nine of seventy-eight firms transferred technology to LDCs in an unaltered form. But transferring the emphasis, this means that nineteen firms *did* alter techniques. In their work, Morley and Smith found "scaling down" modifications by foreign firms in Brazil (see Chapter 11). Hermann von Bertrab studied thirty-one European enterprises in Latin America and forty-two companies in Mexico affiliated with European firms and found that adaptation took place by (1) using older methods, (2) increasing labor-intensity in ancillary functions (e.g., moving materials, packaging, etc.), or (3) redesigning techniques incorporating some older and some modern features.[39] The greatest consideration was to adapt to smaller-scale production. José Giral Barnes has been conducting a program in Mexico for over a decade aimed at reducing the scale of chemical technology to serve smaller markets. The DuPont Company of Mexico has given support to this research, and several of these new technologies have been put into practice by private firms.[40]

Daniel Margolis cites a number of product adjustments undertaken by MNCs to accommodate the LDC consumer market.[41] As might be expected, indigenous firms appear to be more innovative. Since the early 1960s, Mexican steel firms have constructed three new open-hearth steel furnaces, although oxygen converters are considered more efficient in MDCs.[42] Loretta Fairchild's study (presented in Chapter 8) finds

Mexican firms slightly more innovative than those affiliated with U.S. companies. In a previous work to our study, a series of cost-cutting improvements achieved by a Mexican paper producing firm are described.[43] A more extensive survey of the Mexican paper industry, reported by de Márquez in Chapter 9, finds innovative activity by both national and foreign firms. Another case study of a Colombian bakery on the outskirts of Bogotá is instructive. Relying totally on domestic resources, technologies were fashioned to mass-produce biscuits based on traditional recipes.[44] Colombia affords another example in which local firms were induced to adopt technology for producing texturized vegetable protein under the initiative of Colombia's Instituto de Investigación Tecnológica.[45]

Although admittedly an extreme example in view of the ease of substituting between labor and capital, the construction industry in Latin America provides further illustration of the interaction of market forces and the adoption of technologies. W. Paul Strassmann compares the reaction to eleven labor-saving innovations in the construction industry in Peru (where labor unions are strong in construction trades, and wages are relatively high in relation to those of industrial workers) with that in Mexico (where construction unions are less powerful, and wages tend to lag behind those of industrial workers).[46] He found that ten of the eleven innovations had been adopted in Peru, whereas only seven were adopted in Mexico. He concluded that a judicious expansion of construction activity, combined with policies to keep wages from rising too rapidly, could raise employment in LDCs by about 3 percent.[47] The work reported by Strassmann in Chapter 13 of this volume further attests to innovations being adopted by the Latin American construction industry which are extensions of or complementary to the use of traditional building materials. These new methods cut costs while increasing the use of locally available materials and creating more employment.

One of the easiest and most obvious means of adjusting technology to a labor-abundant, smaller market situation is to employ secondhand equipment. Even though careful inspection and appraisal of such equipment is advisable, and repairs, rehabilitation, or major overhauls are sometimes necessary, it is widely used. James estimates that 10 to 20 percent of the capital formation in the form of fixed durable manufacturing equipment in LDCs is accounted for by imported used equipment from MDCs.[48] Constantine Vaitsos reports that such equipment is widely employed in Colombia,[49] as does Strassmann for Puerto Rico and Mexico.[50] Some writers have associated the importation of secondhand equipment with ulterior motives by foreign firms that have nothing to do with the welfare of the host economy.[51] Yet there appear to be well-founded theoretical reasons for the transfer of used machinery,[52] as well as practical ones from the standpoint of the importing LDCs.[53] This is further substantiated by the fact that domestic firms also employ used equipment extensively. Both domestic and foreign firms use it in Brazil.[54] Strassmann found that three-fifths of small manufacturing firms in Puerto Rico and Mexico use secondhand equipment compared to two-fifths of the large firms.[55] Thus, the employment of used machinery appears to be widespread in Latin America as a means of adjusting production techniques to greater compatibility with factor prices and market sizes.

These examples of innovative activity make clear that adaptation to local conditions goes on even in the most discouraging factor markets and under "hothouse" protection from competition. Would foreign and local entrepreneurship remain impervious to a shift in the economic environment increasing the gains from such endeavors? We believe we could reasonably expect greater efficiency in the acquisition and adaptation of technology at both the internal margin (where adaptive firms become more adaptive) and the external margin (where firms initiate adaptive measures). Furthermore, measures to rationalize factor markets so prices of productive inputs more nearly reflect their real marginal contribution to output are available and under the control of domestic Latin American policy.

## Conclusion

To what degree is Latin America restrained by external forces in her policy options for the long-run acquisition of scientific and technical knowledge? Our answer is, quite a bit, but not as greatly as most of the literature, especially *dependentista* literature, leads one to believe. It would appear to us that a shift of internal policy toward a more competitive atmosphere, with factor prices more nearly reflecting true social opportunity cost, and a start toward a more equitable sharing of economic benefits from growth would induce a more rational choice of existing technology, more extensive adaptation to Latin American conditions, and more vigorous efforts to develop new appropriate modes of production. Most of the literature focuses on the fact that there is little effort to adapt technology, especially by MNCs. We agree, yet, as we have seen, adaptation is far from unknown; it does happen to a degree, even under the most unattractive economic incentives. Given this emphasis on what *is* being done rather than what is not, it would be very surprising to us if a substantial enhancement of economic incentives did not result in more firms altering their technology and others stepping up their efforts. We have not concentrated on the precise policies necessary to raise the interest/wage ratio because of the extensive literature dealing with these matters, but clearly these measures lie within the domain of internal Latin American decisions.

One point, however, should be stressed. We are under no illusion that a policy making the internal economies of Latin America more "rational," or uncontrolled, would solve the problem. There may yet be a need for more explicit legislation controlling investment and technology transfer by MNCs, or, more likely, the need to enforce what is already codified. External cooperation is needed if the international patent system is to be overhauled. Also, if Latin American countries are to participate in the very high-cost technologies, the MDCs will have to share their present leadership in these fields. We deny none of these realities, but conclude that there is a range of economic policies subject to internal control that would indirectly improve Latin America's access to and choice of technical knowledge.

We are compelled, nonetheless, to make a concession, and it is potentially a major one. Our analysis concentrates on what we believe *should* be done and why. But much of the dependency analysis concentrates on why policies clearly beneficial to the masses will not be implemented. Those in power, it is said, have been co-opted by foreign influence and have a financial, psychological, and social identification with foreign interests. Controlling Latin American groups maintain the status quo and veto policy changes that would be beneficial to the larger society. We have criticized this view at some length in a previous work, both in terms of content and methodology.[56] It is to a certain extent immaterial, however, whether the source of the obstruction is from outside or within. If the unfolding of future events in Latin America ineluctably indicates a controlling political climate that will not respond to desperately needed socioeconomic changes, technology policies included, we will be forced, however reluctantly, to conclude that more radical paths to change are inescapable.

## Notes

1. U.S., Department of State, *Text of Declaration of Tlatelolco* (Department of State Bulletin 70:1812, 1974), p. 262.

2. Quoted in Alverto M. Piedra, "Problems of Foreign Economic Relations," in *Latin American Foreign Policies: An Analysis*, ed. Harold Eugene Davis and Larman C. Wilson (Baltimore: Johns

Hopkins Press, 1975), p. 37.

3. Ismael Escobar, "Innovation and Development," in Inter-American Development Bank, *The IDB's First Decade and Perspectives for the Future* (Punta del Este, Uruguay: IDB, 1970), p. 90, quoting Richard Adams.

4. Maximo Halty Carrére, "The Process of International Transfer of Technology. Some Comments Regarding Latin America," mimeographed, Pan American Union, Washington, D.C., 1968, p. 3.

5. Bernardo Klicksberg, *Administración subdesarrollo y estrangulamiento tecnológico; introducción al caso latinoamericano* (Buenos Aires: Editorial Paidos, 1973), p. 46.

6. Shiller Thébaud, *Statistics on Science and Technology in Latin America: Experience with UNESCO Pilot Projects 1972–1974* (Paris: UNESCO, 1976), p. 49.

7. A. Nussbaumer, "Financing the Generation of New Science and Technology," in *Science and Technology in Economic Growth,* ed. B. R. Williams (New York: John Wiley and Sons, 1973), p. 176.

8. Montague Yudelman, Gavin Butler, and Ranadev Baneryi, *Technological Change in Agriculture and Employment in Developing Countries* (Paris: Organization for Economic Cooperation and Development, 1971), p. 143.

9. Harrison Brown, "The Role of Science and Technology in Development," in *The Scientific and Technological Gap in Latin America,* ed. Roberto Esquenazi-Mayo, Khem M. Shahani, and Samuel B. Treves (Lincoln, Nebraska: University of Nebraska, Institute for International Studies, 1973), p. 18.

10. United Nations Conference on Trade and Development, "Major Issues in Transfer of Technology to Developing Countries; A Case Study of the Pharmaceutical Industry," mimeographed (1975), p. v.

11. C. D. G. Oldham, C. Freeman, and E. Turkan, "Transfer of Technology in Developing Countries," mimeographed (Geneva: United Nations Conference on Trade and Development, 1967).

12. R. Hal Mason and Frances G. Masson, "Balance of Payments Costs and Conditions of Technology Transfer to Latin America," *Journal of International Business* 5, no. 1 (1974):77.

13. Kenneth J. Arrow, "Economic Welfare and the Allocation of Resources for Invention," in National Bureau of Economic Research, *The Rate and Direction of Inventive Activity* (Princeton, N.J.: Princeton University Press, 1962), p. 615.

14. Charles Cooper, "Science, Technology, and Production in the Underdeveloped Countries: An Introduction," *Journal of Development Studies* 11, no. 2 (1972):11.

15. Raúl Prebisch, "Commercial Policy in the Underdeveloped Countries," *American Economic Review* 49, no. 2 (1959):251-73.

16. For a brief summary of some of the literature critical of the Prebisch thesis, see C. Richard Bath and Dilmus James, "Dependency Analysis of Latin America: Some Criticisms, Some Suggestions," *Latin America Research Review* 11, no. 3 (1976):26-28.

17. Constantine V. Vaitsos, "Transfer of Resources and Preservation of Monopoly Rents," Economic Development Report no. 168, mimeographed (Cambridge, Mass.: Center for International Affairs, Harvard University, 1970).

18. Constantine V. Vaitsos, "Patents Revisited: Their Function in Developing Countries," *Journal of Development Studies* 11, no. 2 (1972):71-97.

19. Grant L. Reuber et al., *Private Foreign Investment in Development* (Oxford, England: Clarendon Press, 1973), pp. 194-95.

20. Wayne Yeoman, "Selection of Production Processes for the Manufacturing Subsidiaries of U.S. Based Multinational Corporations" (D.B.A. thesis, Harvard Business School, Cambridge, Mass., 1968).

21. William H. Courtney and Danny M. Leipziger, *Multinational Corporations in LDC's; the Choice of technology,* AID. Discussion Paper no. 29 (Washington, D.C.: Agency for International Development, 1974), p. 19.

22. Economic Commission for Latin America, "Industrial Development in Latin America," *Economic Bulletin for Latin America* 14, no. 2 (1969):15.

23. Ibid.

24. Tibor Scitovsky, "Prospects for Latin American Industrialization within the Framework of Economic Integration: Bases for Analysis," in *The Process of Industrialization in Latin America* (Guatemala City: Inter-American Development Bank, 1969), p. 39.

25. Economic Commission for Latin America, *Income Distribution in Latin America* (New York: United Nations, 1971), p. 35.

26. Celso Furtado, "The Brazilian 'Model' of Development," in *The Political Economy of Development and Underdevelopment,* ed. Charles Wilbur (New York: Random House, 1973), pp. 297-306, and Osvaldo Sunkel, "Capitalismo Transnacional y Desintegración Nacional," *Estudios Internacionales* 4, no. 16 (1971):3-61; "Big Business and 'Dependencia': A Latin American View," *Foreign Affairs* 50, no. 33 (1972):511-31, and "The Pattern of Latin American Dependence," in *Latin America in the International Economy,* ed. Victor L. Urquidi and Rosemary Thorp (London and Basingstoke: The MacMillan Press, 1973), pp. 3-25.

27. There is scant evidence on this point. When investigating R&D for the chemical, petroleum, and steel industries in the United States, Mansfield found no greater productivity from R&D in larger firms than smaller firms in the petroleum and steel industries. He stressed that his findings were very tentative. (Edwin Mansfield, *Industrial Research and Technological Innovation: An Economic Analysis* [New York: W. W. Norton and Company, 1968], pp. 40–42.)

28. Harrison Brown gauges the "micro-minimum" in basic research within a single university at about ten competent scientists (Brown, "The Role of Science and Technology in Development," p. 24); also, E.A.G. Robinson, speaking of the United Kingdom, claims that "research units of less than ten people were almost purely channels of communication and were not adding to knowledge." Substantial additions to new knowledge are unlikely with research units of much less than 100, Robinson asserts. (E.A.G. Robinson, "Discussion of the Paper by Professor Griliches," in *Science and Technology in Economic Growth*, ed. B. R. Williams [New York: John Wiley and Sons, 1973], p. 88.)

29. National Academy of Sciences, *U.S. International Firms and R. D. & E. in Developing Countries* (Washington, D.C.: NAS, 1973), p. 33.

30. Consejo Nacional de Ciencia y Tecnología, *National Indicative Plan for Science and Technology* (Mexico, D.F.: CONACYT, 1976).

31. Awesome though the technology gap between Latin America and the MDCs undoubtedly is, there are several factors which at least reduce the width of the chasm. First, much R&D in MDCs is directed toward military-connected pursuits. Second, of the economically motivated research in MDCs, a great deal of effort goes into product development—much of which entails minor changes for purposes of differentiating the product in imperfectly competitive markets. Third, conceptually, costs of innovation should be included in R&D expenditures, but, in practice, they are seldom picked up fully in the data. These points should not be taken lightly. In the mid-1960s "economically motivated" R&D was considerably below total R&D expenditures for the United States, Western European countries, and Japan. Furthermore, in the United States, as much as 90 percent of industrial R&D is aimed at product innovation. As we will see later in this chapter, a review of case studies reveals that significant innovational activity does take place in Latin America, leading at least to the working hypothesis that, if properly measured, the expenditure on innovation as a percent of GNP would be less disproportionate between Latin America and MDCs than R&D differentials. (R. C. O. Mathews, "The Contribution of Science and Technology to Economic Development," in *Science and Technology in*

*Economic Growth,* ed. B. R. Williams [New York: John Wiley and Sons, 1973], pp. 7, 12.)

32. Flavia Derossi, *The Mexican Entrepreneur* (Paris: Organization for Economic Cooperation and Development, 1971).

33. David R. Thompson, "Imported Technology and National Interests in Latin America" (Senior Seminar in Foreign Policy, 16th sess., U.S. Department of State, 1973–74), p. 19.

34. Olga Pellicer de Brody, "Mexico in the 1970s and Its Relations with the United States," in *Latin America and the United States: The Changing Political Realities,* ed. Julio Cotler and Richard R. Fagen (Stanford, California: Stanford University Press, 1974), p. 331.

35. Consejo Nacional de Ciencia y Tecnología, *National Plan for Science and Technology,* p. 208.

36. Richard D. Morgenstern and Ronald E. Muller, "Multinational Versus Local Corporations in LDCs: An Econometric Analysis of Export Performance in Latin America," *Southern Economic Journal* 42, no. 3 (1976):399–406.

37. Ian Little, Tibor Scitovsky, and Maurice Scott, *Industry and Trade in Some Developing Countries: A Comparative Study* (London, New York, and Toronto: Oxford University Press, 1970), p. 174.

38. Bela Balassa et al., *The Structure of Protection in Developing Countries* (Baltimore and London: Johns Hopkins Press, 1971), p. 56.

39. Hermann von Bertrab, "The Transfer of Technology: A Case Study of European Private Enterprise Having Operations in Latin America with Special Emphasis on Mexico" (Ph.D. diss., University of Texas at Austin, 1968).

40. José Giral Barnes, "Development of Appropriate Chemical Technology: A Programme in Mexico," in Organization for Economic Cooperation and Development, *Choice of Technology in Developing Countries, An Overview of Major Policy Issues* (Paris: OECD, 1974), pp. 182–86.

41. Daniel Margolis, "Multinational Corporations and Adaptive Research for Developing Countries," in *Appropriate Technologies for International Development: Preliminary Survey of Research Activities* (Washington, D.C.: Agency for International Development, 1972).

42. Economic Commission for Latin America, "Industrial Development in Latin America," p. 52.

43. Dilmus D. James, "Used Automated Plants in Less Developed Countries; A Case Study of a Mexican Firm," *Inter-American Economic Affairs* 27, no. 1 (1973):31–46.

44. Daniel Schlesinger and Lucia de Schlesinger, "Mass Production of Cakes in Colombia," in International Labor Office, *Automation in*

*Developing Countries* (Geneva: ILO, 1972), pp. 131-46.

45. Diógenes Hill, Harvey Palaéz, and Héctor Botero, "Análisis preliminar de un fenómeno de innovación tecnológica de interés para el país: Introducción de las proteínas vegetales texturizadas en el mercado nacional" (Bogotá: COLCIENCIAS, 1976).

46. W. Paul Strassmann, "Construction Productivity and Employment Objectives in Developing Countries," *International Labour Review* 101, no. 4 (1970):503-18.

47. Ibid., p. 516.

48. Dilmus D. James, *Used Machinery and Economic Development,* MSU International Business and Economic Studies, Division of Research, Graduate School of Business Administration (East Lansing, Michigan: Michigan State University, 1974).

49. Vaitsos, "Transfer of Resources and Preservation of Monopoly Rents," p. 38.

50. W. Paul Strassmann, *Technological Change and Economic Development; The Manufacturing Experience of Mexico and Puerto Rico* (Ithaca, New York: Cornell University Press, 1968), pp. 207-09.

51. James D. Cockroft, "Mexico," in *Latin America: The Struggle with Dependency and Beyond,* ed. Ronald H. Chilcote and Joel C. Edelstein (New York: Halstead Press, 1976), p. 281; Ignacy Sachs, "Selection of Techniques: Problems and Policies for Latin America," *Economic Bulletin for Latin America* 15, no. 1 (1970):20; Vaitsos, "Transfer of Resources and Preservation of Monopoly Rents," p. 38.

52. Amartya Kumar Sen, "On the Usefulness of Used Machines," *Review of Economics and Statistics* 44, no. 3 (1962):261-78; M.A.M. Smith, "International Trade in Second-hand Machines," *Journal of Development Economics* 1, no. 3 (1974):262-64; Sandra L. Schwartz, "Second-hand Machinery in Development, or How to Recognize a Bargain," *Journal of Development Studies* 9, no. 4 (1973):545-55.

53. Dilmus D. James, "Used Automated Plants in Less Developed Countries, a Case Study of a Mexican Firm"; *Used Machinery and Economic Development*; and "Second-Hand Machinery in Development: Comment," *Journal of Development Studies* 11, no. 3 (1975): 230-33.

54. Courtney and Leipziger, *Multinational Corporations in LDCs,* p. 19, citing Leff.

55. W. Paul Strassman, *Technological Change and Economic Development,* pp. 208-09.

56. C. Richard Bath and Dilmus James, "Dependency Analysis of Latin America."

# 3
# On the Diffusion of Technology in Latin America

*David Felix*

This chapter on Latin American experience with technology diffusion concentrates on the period 1860 to the present. At the beginning of that period most of the present national frontiers had been defined (apart from later seizures at the expense of Bolivia, Ecuador, and Colombia), and when the post-independence political chaos had in most countries settled down to a "normal level of instability."[1] It was an era when mounting food and raw material needs of industrializing Europe and cheapening ocean transport costs began rapidly to augment overseas trade and investment opportunities. It was also a period when European migration, impelled by agricultural transformations and rising demographic growth, and induced by cheaper ocean transport, began accelerating.[2] The period has thus been marked by an expanding Latin American involvement in foreign trade, investment, and immigration and by an inflow of European and, later, U.S. technology, which partially transformed the Latin American productive base.

Since the Napoleonic Wars, Latin American countries have been, except for occasional interludes, wide open to foreign investment and the importation of foreign technology. Until the 1929 depression they were also quasi-free traders, with tariffs mainly imposed for revenue purposes. Protectionist rumblings from incipient industrial groups began to be heard by the turn of the century, but the efforts did not take widely (as they had earlier in continental Europe and the United States) until the economic crisis of the 1930s. Prior to the 1860s, the readiness of the Latin Americans was not matched by the willingness of the Europeans. But after that date there was consummation, reaching orgiastic proportions during the intermittent bonanza periods that have

This chapter has been adapted from a longer monograph presented at the Conference on Diffusion of Technology and Development, Bellagio, Italy, April 21-26, 1973.

marked Latin American economic history since the mid-nineteenth century. To explain Latin American dualism and the limited diffusion of advanced technology, we have to look for something other than barriers to the free working of international market rules of the game.

For this purpose the 1860–1970 span can usefully be divided into Period I, running from the 1860s to the 1920s, and Period II, running from the 1930s to the present. The division is justified partly by major shifts in trade and domestic policies, which had their onset with the Great Depression in the larger Latin American countries. It is also justified because the gradual shift since the late eighteenth century in the main sources of innovation, from empiricist tinkering and learning-by-doing to organized science-technology efforts, had reached the point after World War I where the socioeconomic conditions for becoming a technologically progressive society had changed dramatically.

The main thesis of this chapter is that Latin American countries, although eager borrowers of imported technologies, have been institutionally out of phase in both periods, each time for a somewhat different set of reasons. They have been unable to creatively adapt and diffuse advanced technologies on a broad enough scale to become progressive monistic economies. In Period I, when most knowledge, risk, and capital requirements for adapting and diffusing new technology were still within the capabilities of alert artisans and individual entrepreneurs, Latin American countries were deficient in both. Out of the painful import substitution industrialization (ISI) efforts of Period II, an adequate base has emerged for effectively elaborating on and diffusing nineteenth century technology (at least in the more industrialized Latin American countries). But technology has also moved on, and the institutional requirements for broad-scale diffusion of mid-twentieth century technology remain out of reach. Hence, Latin American countries and the majority of their citizenry continue to suffer the many socioeconomic ills that result from excessive dependence on imported technology.

## The Technological Frontier and How It Grew

In the nineteenth century, the search for new technological solutions was primarily guided by work experience and only vaguely modified by formal scientific insights, whereas in the mid-twentieth century such technological search has been drawing largely on the formal search rules provided by close intellectual linkages between science and technology. In the first half of the nineteenth century, chemistry was just emerging from its phlogiston phase; electromechanics was still a recondite laboratory game. At the same time, the intellectual gulf between theoretical and practical metallurgy or between the formal rules of engineering and the rules of thumb used in machine designing and building was almost as great as that between constructing nonlinear, multisector, multiperiod, optimizing planning models and drawing up and implementing macroeconomic plans. How much influence the prestige of Newtonian mechanics exercised on the mind-set of seekers after lucrative technological solutions is a matter of dispute, but it is evident that two major lapses prevented a closer relationship. One was the infeasibility of applying the scientific and engineering rules useful for theoretical relationships under simple, ideal conditions to the messily complex and varied circumstances encountered in production. The other was that the ability of machine builders to provide the fine tolerances, speeds, and control devices needed to make such rules operative was limited. Gradually, over the century, the two gaps were filled by advances in hydraulics, metallurgy, chemistry, and electromechanics and by counterpart improvements in machine building. But through most of the nineteenth century, the frequently proffered advice of successful engineers and inventors to aspiring young careerists to apprentice themselves early in practical activities rather than pursue advanced scientific or engineering education retained much of its initial plausibility.

Paralleling the trend toward formalized technological knowledge and search methodologies have been changes in the relationship between artisan and factory products. For most of the nineteenth century, the quality of factory products was generally below that of well-made artisan products. They sold because they were cheaper than, while competitive in quality with, products from the lower half of the artisan quality range. Marketing was therefore mainly a matter of pricing low and of shifting most of the distribution, selling, and sales finance to independent intermediaries. Artisans producing in the upper quality range survived and even flourished. Their prosperity was sustained by the large "discretionary" income of the upper classes and of the rising middle classes. By the twentieth century the quality relationship had reversed itself. Aided by advances in metallurgy, chemistry, and machine building, factory products have surpassed artisan products in quality as well as in cheapness for most products.

Broad technological diffusion and economic growth in the nineteenth century thus required a different set of human capital and behavioral conditions than has been prevalent in the twentieth century. Whether growth and technological progress reinforced or dampened initial socioeconomic dualism depended in each country on the extent to which the conditions were present, either as a pre-Industrial Revolution social inheritance or as a consequence of national developmental policies.

The essential early nineteenth century requirements for broad-based capitalist growth were: a rich layering of specialized artisanry, with the upper layers at least literate and *au courant* with innovations in artisan technology; widely distributed ownership of the main preindustrial asset, land; a well-developed set of interconnected internal markets, transport facilities, and mercantile intermediaries, mainly linking agriculture with artisan industry within geographic subregions, but also generating some interregional and international trade. With a rich endowment of these conditions, technological innovations connected with artisan ingenuity and financial and marketing innovations emerging from commerce and trade could provide the technological and capital accumulation basis for broadly diffused technical progress and growth, relying primarily on private initiative.

No early nineteenth century country was fully endowed with all these characteristics, but some were more richly endowed than others—so much so that they visibly drew apart from the rest, forming by mid-century the core of today's economically developed club. The rest, whether independent or colonial dependencies, were coalescing into the backward outer world, heavily dependent on technological and capital spillovers from the club's kitchen.

Nineteenth century technological progress in the advanced group was highlighted by periodic spectacular discoveries, but most of it was a steady accretion of small improvements. The latter were modifications of technological discoveries to solve application difficulties, spillovers of solutions from one activity to another, and the result of recombining known technologies. The accretions would, from time to time, lead to new "spectaculars." Although Britain was the industrial workshop of the group, dazzling discoveries and accretions also emanated from the other members. The technological borrowing within the group was a somewhat unbalanced interchange rather than a one-way flow from leader to followers. But since technological exporting is the most advanced manifestation of technological prowess, its presence in all countries of the group is a prime indicator that there were only moderate differences in the endowment of the various group members for adopting and diffusing technology. In the lead country, Great Britain, the process took place almost entirely through private ingenuity and entrepreneurship guided by market forces. In the latecomers, there was more modifying and supplementing of market forces by state action: through protection, state technical institutes, and investment in infrastructure. By twentieth century norms, the role of the state was, however, quite modest.

Late twentieth century technological progress in the advanced-country group continues to be composed of intermittent "spectaculars" amidst a myriad of small accretions, but the sources have undergone a transformation. Unschooled empiricism has been largely displaced by research and testing laboratories, pilot plants, computer simulation, highly specialized scientific and technological communications media, and other expensive paraphernalia of modern innovating. The clever artisan has been largely displaced by the highly educated "knowledge industry." Today's less developed country (LDC) trying to cop a ride on the technology train is confronted by a quite different and more rapidly moving vehicle than its nineteenth century counterpart.

The empiricism of nineteenth century innovating and the consumer demand trends favoring artisan products had also generated a fairly close link between the rising shares of output and industrial employment. In the twentieth century, the link has been greatly weakened, the industrial output share growing at a faster rate than in the nineteenth century, while industrial employment has remained virtually constant. The contrast is even sharper in agriculture, where twentieth century technological spectaculars in agricultural machinery and chemistry have sparked an absolute decline of employment in advanced countries, concomitant with an acceleration of output. Twentieth century growth has been channeling a continually increasing proportion of the labor forces of advanced countries into tertiary employment—finance, sales promotion, and distribution, the "knowledge industry," civilian bureaucracies, and the military—whereas nineteenth century growth channeled it mainly to industry, construction, and the newly settled overseas regions.

It is conventional to explain the labor-saving trends in industrial and agricultural technology as economizing responses to labor scarcity and rising wages. But if we see technological choice as involving creation, rather than mere activation of unused blueprints of inventions already made, other dynamic scenarios also come into view. Most notable is the search for new ways of exploiting economy-of-scale dynamics. The profitability of enlarging plant scale resides in widely applicable physical principles. Effectively exploiting these principles required search for improved structural materials and design innovations, improvements in power plants, pumps, machinery, switching, and control devices. The increase in downtime loss for larger plants from adulterated inputs and machine breakdown as a result of human error also impelled search for "automating" quality control and machine guidance. All this, in turn, has required innovations in distribution, goods promotion, and taste molding (along Galbraithian lines) to move the larger volumes of output smoothly through the marketplace. Overall, it has been a process of displacing raw labor by fixed capital and skilled maintenance, marketing, and administrative labor, and of progressive increases in plant size.

The ample rewards, including increased monopoly power, from exploiting these physical principles lend plausibility to the economy-of-scale scenario, which implies, of course, a different dynamic interaction between wages and productivity than the labor-scarcity case. Instead of having increases in capital intensity be impelled primarily by rising wages, they are a by-product of the search for profitable new economies-of-scale, with increased profits activating wage pressures through other social mechanisms than scarcity.[3]

Put thus baldly, it becomes clear that neither scenario can explain, by itself, the complexities of the labor market. But the economy-of-scale scenario does help account for major anomalies in the labor-scarcity scenario, such as rising wages, despite the infrequent incidence of full employment in advanced economies prior to World War II.[4]

Each scenario also has a quite different implication for LDC technological borrowing. If the trend toward increasing capital and skill intensity has been impelled primarily by labor scarcity, then, in their search for factor-appropriate technology, the LCDs need go no further than to activate techniques discarded by the advanced countries. On the other hand, the more widely relevant the economy-of-scale scenario, the larger is the proportion of discarded technologies that are absolutely inefficient, regardless of wage rates. In that case, LDCs would have to learn to creatively modify and adapt the discarded technologies fairly early in their development in order to minimize the social costs of their inefficiency and to broaden technological diffusion.

## The Diffusion of Technology in Latin America—Period I

Against the moving backdrop of advanced-country technology, let us look at the Latin American experience during Period I (1860–1920s). Since Latin America is a fairly diverse region, some subdivision is necessary, even for general observations, according to the particular aspect of the experience in question.

A common feature around 1860 was an inherited structure of isolated village and subregional markets, with very modest levels of inter-regional and international trade. Village trade, particularly in the Cordillera Indian regions, was heavily imbued with direct bartering of peasant foodstuffs for artisan-made implements and household goods, much of it also the part-time products of peasant artisans. A modest layering of merchant, financial, and transport intermediaries was related primarily to village-city trade and interregional and international trading.

Provincial cities and towns made up a smaller percentage of urban population, and village-city and interregional trade a smaller proportion of total trade than in eighteenth century Western European countries or Tokugawa Japan. The sparsity of provincial towns reflected the high concentration of landownership in Latin America and the semisubsistence character of peasant life. In addition, urban encephalitis was already visible in the mid-nineteenth century, a very high proportion of the small urban population residing in the national capital and major port of each country, where their consumption needs, apart from foodstuffs, tended to be met in large degree by imports. Trade between regions, including intercountry exchange, consisted mainly of noncompeting agricultural and mineral-based goods in which the exporting region usually had some overwhelming natural advantage: copper from central Chile, hides and salt beef from the Argentine littoral, coffee from Brazil's Paraiba valley, sugar from northeast Brazil, alpaca wool from Peru, *yerba maté* from Paraguay. There was some long-distance trade in artisan products of indifferent quality by European or Asiatic standards, but much of it was a stagnating remnant of colonial specialization between silver mining areas and food, clothing, and implement-producing towns. In general, the trading patterns resembled feudal rather than eighteenth-century Europe.

Overseas trade, on the other hand, had distinctive features, most notably a wide range of varied imports of tools and consumer manufactures. The latter consisted of factory products—e.g., textiles and ironware—along with luxury artisan goods. Coastal ports and capital cities, where even wealthy landowners resided during much of the year, were the main markets for imports. Some, including smuggled goods, also entered the provincial marketing networks, but these were secondary markets for such goods. Latin America's persisting socioeconomic dualism was thus already well reflected in its contrasting internal and overseas trading patterns.

A second distinctive feature was the substantial involvement of foreign merchants and bankers in the overseas trade. At mid-century these were almost entirely British houses. Later they encountered increasing foreign competition from firms of U.S., German, and French origin, but jointly, throughout Period I, foreign firms dominated the mercantile, financial, and transport intermediation of Latin American overseas trade. Exports, limited to a small number of natural resource products that traveled well, were also handled mainly by the foreign intermediaries. The range of exports widened during the subsequent half-century, but did not extend beyond raw or partially processed primary products.

The economic dualism was rooted in the highly unequal distribution of capital assets (notably, land), in other institutional inequities, and in the uneven incidence of exploitable natural resources and transport obstacles. Developmental policies, however, seemed to concentrate mainly on transport barriers, and this led to the first major influx of advanced nineteenth century technology, the railroad. Short experimental lines, mainly government-financed, were constructed in the 1850s and 1860s, with the major outburst of railroadization beginning in the 1870s. It peaked by the end of the century, but did not markedly abate before World War I. The technological lag with the United States and Western Europe in railroadization was thus about three decades, and with Great Britain only about four decades.

*The Limited Spread of Technology*

The general developmental effects of railroadization in Latin America, including technological spillovers, were much less numerous than in the United States and the Western European countries. There were a number of reasons for this anomaly. One was that railroad building, by narrowly following paths indicated by emerging export opportunities, resulted frequently in unconnected lines, each linking an interior area with the nearest accessible port. This was especially the case for lines constructed and financed purely by private initiative, but was also a feature of the many lines receiving state subsidies. Later efforts to weave lines into national networks were hindered by the fact that the initial ones had been built with varying track gauges, a rather unnecessary replication of early British experience. (If officials had borrowed from nineteenth century U.S., French, or German transport planning schemes, this situation might have been avoided.) Apart from the obstacles to integration from gauge differences and from the high cost of building transverse linking lines across mountainous and jungle terrain, the political will to incur the requisite costs was also weak. Although most of the trackage was, in fact, built with some tax subsidies, risk underwriting, or direct financing by the state, the state was controlled by a landowning class whose main interest was in exportation and the appreciation of land values—not in slow-yielding public investment to encourage internal markets and regional growth poles through a broadly linked national network. Thus, Argentina, whose flat terrain was suitable for low-cost railroad building, nevertheless developed a radial pattern originating mainly in Buenos Aires and, secondarily, in the up-river ocean port of Rosario. Consequently, colonial artisan centers like Tucumán, Salta, and Córdoba could connect with markets in expanding wine- and fruit-growing Andean regions a few hundred miles away only by means of 1,000- to 1,500-mile circuitous routings.

A second element reducing the general developmental effects of railroadization in Latin America was the limited amount of technological spillover. Local repair shops were established in most Latin American countries, but rails, rolling stock, and most ancillary equipment continued to be imported well into Period II.

The limited technological spillover was the result of a number of factors. One was the maturity and relative standardization of the infrastructure technologies and their embedment in a well-developed overseas network for equipment and supplies, design procedures, and construction firms who had ample access to the financial markets of the advanced countries at the time the technologies began to be disseminated to Latin America. Adaptation of these technologies to Latin American conditions was therefore reduced to solving terrain problems; arranging for reliable supplies of local labor, bulky building materials, and foreign supervisory labor; importing equipment and spare parts; coordinating local with foreign financing; and threading through legal and political idiosyncrasies. The limited technological spillover thus resided, in part, in a Hirschman-like paradox, in which the immediate benefits to Latin American countries from importing mature technologies were partly offset by a diminished need for the sort of local technological and organizational experimentation that had impelled spread effects in the advanced countries. Tying arrangements between the foreign-run infrastructure firms in Latin America and the overseas equipment manufacturers reinforced the situation. In advanced countries, such monopolizing arrangements mainly resulted in intraeconomy profit transfers, along with some domestic resource misallocation. In Latin American countries, they raised the foreign exchange costs and retarded the growth of local content in constructing and operating the infrastructure. The tying problem persisted well into Period II—even after competing motor transport, devaluation, and foreign exchange rationing were seriously undermining railroad profits, foreign-owned railroads persisted in their tying arrangements with overseas suppliers. It took nationalization of the railroads and public utilities to free Latin American countries to seek cheaper overseas sources of supply and to accelerate the growth of local content.

Neither the tying schemes nor the high import content would have endured so long, however, had the initially deficient base of industrial and entrepreneurial skills improved during Period I at an adequate pace. The moderate growth of labor skills did not raise the opportunity cost of tying contracts (and, more generally, of relying so heavily on imports) to the point where the infrastructural firms began to find it more profitable to seek local suppliers. Neither were there many eager local entrepreneurs in Period I beating against the barriers to local supplying. Additionally, public policy during this period was not especially oriented toward augmenting industrial skills and entrepreneurship or breaking down monopolistic entry barriers. The heart of the explanation for the limited technological spillovers doubtless resides in these three internal deficiencies of Latin American countries during Period I.

*Local Industry*

The initial artisan base in Latin America was broadly divided between a native crafts sector and a "modern" monetized sector. The former, embedded in the oppressed, introverted Cordillera Indian cultures, was too distant from European technological and institutional modes to have been a fertile seedbed for technological spillovers. Careful nurturing through enlightened state policies might have changed this, but Period I policies continued to reinforce the colonial practice of treating the Indian villages primarily as sources of agricultural and mining labor. The main difference during Period I was the partial substitution of legal chicanery and market power, notably an accelerated takeover of Indian village land and an increased resort to debt servitude for colonial *force majeure*. The result was degeneration of Indian crafts rather than the progressive linking of Indian with European artisan skills.

The "modern" artisan sector was oriented primarily to supplying implements for mining, agriculture, construction, and road transport as well as basic artifacts such as processed food, textiles, furniture, and kitchenware for the strata of monetized households. Since the quality was usually consonant with the lower levels of European artisanry, the "modern" artisan products were vulnerable to the rapidly rising import competition from cheaper factory products which export bonanzas, improved transport infrastructure, and rising urban and landowner incomes made possible. The flood of competitive imports proved too great for the local artisanry to adapt through learning adjustments. Adaptation required continual technological and organizational improvements, for which a necessary condition was an adequate rate of capital accumulation and external credits to finance the improvements.

Artisanry was left largely to shift for itself, which meant cutting prices and sweating labor in an effort to survive. The vicious circle of declining profits, poor credit risk, inability to finance improvements, and inability to upgrade quality and productivity thus led to the decay of many artisan lines.

To be sure, new lines were also stimulated by rising income and the opportunities to service export, urban, and infrastructural expansion. The ability of new lines to dampen dualistic tendencies was constrained, however, by two additional features of Period I. One was the preference for imported goods that accompanied the growth of middle- and upper-class income. French furniture, architecture, and tableware, Italian statuary, British and French clothing, and European liquors were some of the more public manifestations of this consumption trend. The second was liberal immigration policies to attract European labor—in Brazil and the River Plate countries these included use of recruiting officers, propaganda campaigns, and subsidized passages. The subsidies were primarily for agricultural labor, but many who came never left the port of entry or soon fled harsh rural work for the big city. In Brazil and the River Plate countries, European immigration was a flood; in the others, it was more a trickle. But in all countries, the immigrants supplied much of the skilled labor and entrepreneurship for the new industrial artisan lines. Both the consumption and the labor supply trends dampened the expansion of opportunities for native artisans, and for the ruling elites, the need for measures to upgrade native artisan skills and enterprise.

### *Export Agriculture*

In the export sectors, entrepreneurial and technological trends offer a somewhat different picture. While foreign entrepreneurs, mainly immigrants, were important in export agriculture during the course of Period I expansion, most of it outside of Central America and the Caribbean remained in native hands. This was even more the case for *hacienda* agriculture, producing mainly for the domestic urban markets. Large estates were frequently passed along within the same families from colonial times, but in many countries the bulk of what in the twentieth century came to be called the landed oligarchy was of nineteenth century origin. Entry into large commercial agriculture was fairly open to individuals and groups with capital and/or political influence—immigrant farmers and agricultural technicians, urban merchants and lawyers, politicians and military officers. Some did it through astute land buying, others dispossessed small holders by assiduously searching for ambiguous colonial titles and gaining possession through the courts (as in Colombia) or through seizure for accumulated debts (as in Mexico). In Argentina, the successful Indian wars of the 1870s liberated huge tracts of humid pampa lands which were then sold by the state in large tracts or awarded to military officers in payment for war service. Some export agriculture also emerged from European colonization schemes, with initial funding by overseas religious or charitable institutions. What did not occur was substantial rural upward mobility; peasants did not become farmers, much less *hacendados.* Instead, rural bimodalism was accentuated in most countries: semi-subsistence minifundia and commercial latifundia, with not much in between. There were a few subregional cases (e.g., on the western spurs of the Colombian Andes, where the producers of the major commercial crop—coffee—were mainly small holders), but these were exceptions to the general rule.

New agricultural regions lacked a reserve agricultural army. Labor-hungry *hacendados* were forced, therefore, to draw labor away from settled regions or to attract foreign labor. Typically, they opted for the second alternative, partly because European labor was considered more skilled and disciplined, but also probably to avoid sociopolitical consequences of upsetting labor arrangements in the settled regions. Since individual costs of recruiting foreign labor were unattractively high, the state was pressed into acting as a foreign recruiter. Thus, the countries that in Period I were undergoing major agricultural expansion in sparsely populated subregions—notably Brazil in the state of São Paulo, Argentina in its littoral provinces, and Uruguay—undertook major promotion and subsidization of European immigration in the late nineteenth century. For Brazil the effort had been forced by slave emancipation in 1888.

In the newly settled regions, *hacendados* replicated many of the cash-economizing elements of the labor system prevailing in the settled regions, albeit with modifications forced on them by local circumstances. On Brazilian coffee *fazendas,* labor would be allowed initially to clear forest and brushland for their slash-burn food raising; then, when the coffee trees planted concurrently reached maturity, they would be given more permanent food plots in return for coffee labor. European labor, however, did not take well to the primitive conditions, turnover was high, and there were embarassing exposés in the home countries, so by the twentieth century *fazendeiros* were finding it more economic to bring in labor from the depressed states of Minas Gerais and the northeast. Argentina evolved a short-lease tenancy system that suited the field rotation requirements of the mixed cattle-grain agriculture that had sprung up on the humid pampas. Tenants raised and harvested wheat or linseed for a couple of seasons, usually with their own draught animals and implements, but aided by migratory European and native labor during harvest. They then planted alfalfa and left to take over another enclosed field, either on the same *estancia* or on another. The *estanciero* shared in the grain income and had the continual supply of fresh alfalfa pasture that was essential for raising export-quality beef for the European market. The system also required a permanent crew of estate laborers for cattle raising and estate maintenance; they were paid partly in cash as well as in kind. The Argentine system was also more dependent on purchased inputs—barbed wire, jute bags, etc.—than the traditional *hacienda,* but the short-lease tenancy system greatly reduced cash-flow needs and gave the Argentine *estancia* many of the risk-averting benefits of the traditional system.

In commercial agriculture, technological search and diffusion mainly consisted of selecting and adapting imported seeds, breeding stock, fruits, and vines. Virtually all of it was done under private aegis, large estates taking the lead in the case of cattle and coffee, and European colonists playing the lead role in horticulture. European colonists were also instrumental in starting commercial wheat cultivation in Argentina. Later, wheat became a major component of cattle-grain *estancia* agriculture, but in 1870, when wheat was introduced, *estancieros* were reportedly very negative toward the grain, changing their minds only when the colonists demonstrated its profit potential. Similarly, Welsh colonists were the main initiators of sheep cultivation in Patagonia.

Individual efforts were supplemented by agricultural societies and were guided indirectly by the pricing and grading practices of the large processing and export marketing firms. Yet, while modifications were bred into imported breeds and plants over the years, there were few exportable innovations. About the only important exception was bananas—varieties of standardized quality that also traveled well were needed for overseas exporting. These were developed through systematic research by large U.S.-owned agrobusinesses that had set up in the "banana republics" of Central America in the early twentieth century. But overall, Latin America contributed far less to international agricultural technology during Period I than its Indian cultures had done four centuries earlier.

Despite a basic dependence on agriculture, there was little direct government participation during Period I in agricultural research or the dissemination of agricultural knowledge. Some of the countries set up agronomy schools on a small scale, but government research stations and exte[illegible] services are mainly a post–World War II development. It is unlikely that the surprising lag was due to an unperceived need for collective channels to disseminate technological information on the part of commercial farmers. Indeed, well before the turn of the century, agricultural societies with regional branches and product subdivisions appeared in most countries, assuming an important role in disseminating technical and marketing information. They sponsored fairs, published farm journals, promoted standards, and acted as the organized political spokesmen for large farmers. Membership tended to be restricted to large holders and came to be considered a major status symbol for wealthy *arrivistes;* so much so that social critics came to view the societies as the very embodiment of the landed oligarchy. However, the agricultural societies, their information geared to the needs of large agriculture, were judged by the latter to be adequate supplements to individual search efforts, the main deprivants being the *minifundistas.* A far-sighted government might have seen the need for action to bring up the techniques of this heavily populated depressed sector, but the political situation was not ripe for such enlightenment. The *latifundistas* were politically powerful but uninterested; the *minifundistas* were politically impotent and unaware.

Agricultural expansion also required major improvements in marketing and, in some cases, technological breakthroughs in processing. Most of the improvements and advances were of foreign origin, and the export marketing and processing sectors in Period I tended to have a predominance of immigrant or overseas ownership. Tuned into the quality requirements of overseas markets, and faced with delivery and perishability problems, the processing and marketing firms used their buying power to pressure farmers into adapting cattle and produce to their quality needs. They were instrumental in promoting many institutional innovations in marketing and in importing new techniques for packaging and storing produce. The expansion of the main urban centers, the influx of immigrants, and the spread of European tastes in food among the wealthier strata of the urban population also encouraged the growth of food-processing firms oriented to the expanding urban markets. Immigrant capital and entrepreneurship were major instigators of the growth of breweries, dairy plants, flour mills, sausage makers, bakeries, and the like. In Buenos Aires, census data indicate that immigrant firms dominated such activities, but impressionistic evidence suggests a greater participation of native capital in some of the other countries. In any event, the specialized input requirements of these food processors further helped guide the growth of agricultural marketing institutions and the choice of seed and breed in agriculture.

*Mining Development*

In mining, which was concentrated in the *cordillera* countries, the development dynamics were rather different. Native capital pioneered in the early expansion of mineral exporting and was then displaced by large-scale foreign enterprises. Mining in many of the *cordillera* countries had been the core of the export sector during the colonial era, and around it had developed a largely indigenous set of skills, technology, and supplier networks. After independence, silver exports stagnated, but exports of other non-ferrous minerals (notably copper, lead, nitrates, and tin) expanded, particularly after mid-century. Native-owned mines were generally small, worked rich, shallow ore beds, employed simple, labor-intensive techniques, and tended to be technologically static. As world demand expanded, freight costs fell, and overseas mining technology advanced, Latin America came to be a prime region for foreign mining investment. By the eve of World War I, export mining had become overwhelmingly a foreign-controlled activity. With the transition, input linkages with the host economy were progressively reduced. The new technologies were heavily labor saving, depended on foreign equipment, fuels, and technical and supervisory personnel, transferred profits to foreign owners, and paid minimal local taxes. A 1920 official estimate for Chile gives the retained Chilean share of domestic copper sales as 11 percent.[5] The changeover transformed the mining sectors into disconnected enclaves of the host economies.

The rapid displacement of native entrepreneurship in mining and its limited participation in new productive activities outside of agriculture has long been a topic of speculative analysis and social criticism. No doubt the diffusion of technology would have been greater and dualistic trends less prevalent had native entrepreneurs behaved like Yankees or Victorians rather than Latinos. Why didn't they? Most answers divide into those emphasizing sociopsychological factors and those stressing market imperfections. Constant repetition of the question has not brought matters closer to consensus. Earlier writers, innocent of proper social science methodology, tended to mix sociopsychological traits indiscriminately with market imperfections in their answers. Subsequent advances in economic and sociological sophistication have led to sharpened reformulations of the questions that have, in turn, tended to confuse matters further. The question remains too important for understanding the limitations of the market as a mechanism for diffusing technology to pass over in complete silence. We therefore cite three major cases in which foreign capital displaced local capital in order to illustrate some of the complexities of the question, and then suggest a few generalizations on the entrepreneurial question.

*The Argentine Meat Trade*

The first case is the River Plate export meat trade. Its origin dates back to the early nineteenth century, when *saladeros* (beef drying and salting plants) were established along the shores of the River Plate and its tributaries. The *saladeros,* which processed the abundant *criollo* cattle, offspring of escaped animals brought over by Spanish colonists, were financed by *estanciero* and merchant capital. They developed a thriving export market, primarily in Brazil and Cuba, where jerked beef was an important slave food. The major technological breakthrough that transformed pampa ranching came in the 1870s, when innovations in freezing and insulating were applied to the ship. With the necessary improvements, mainly British, substantial exporting of frozen Argentine and Uruguayan mutton and beef began in the 1890s, and after further advances in temperature control, the famous Anglo-Argentine chilled beef trade came on the scene around 1900. The *saladeros,* on the decline, were replaced by *frigoríficos.* On the *estancia* British breeds, shorthorns, Herefords, and later, Aberdeen Angus, replaced the tough, slow-maturing *criollo* cattle. Argentine capital was involved in both transformations but, in the case of the export *frigoríficos,* dropped out after the pioneering phase. By 1900, the export *frigoríficos* were virtually all British; Argentine packers retreated to service the inland urban market.

Why did Argentine packers pull out of the export trade? Not because of major barriers to selling in the British market (other than scale) or because profits in export meat packing were low. The opposite is indicated by the intermittent warfare, interspersed with long periods of stable cartelization, that characterized the meat export business during its bonanza era. Warfare would break out with efforts by large American houses—Swift, Armour, and Wilson—to break in or expand their share. Settlement involved formal, precise reallocation of market shares between the Anglo and American *frigoríficos.* The ability of the American firms to penetrate more and more deeply into the British market indicates that there were no very difficult tie-ins or other peculiarities to bar foreign marketeers. The long periods of stable market sharing imply a degree of market power that could not help but be used to set high profit margins. Indeed, the post–World War I history of Argentine meat exporting is replete with complaints against the "meat trust." This led to the establishment by the ranchers of a cooperatively financed and state-sponsored rival export *frigorífico* in the late 1920s.

The most plausible answer is, therefore, the scale barrier, and it is perhaps more an organizational than a financial one. Meat packing for the home market, which could begin on a modest scale and grow in organizational and technological complexity at an incremental pace, was merely another manifestation of nineteenth century technological absorption. Competing in export meat packing, which required major organization and technological leaps, partly in processing but even more in marketing, was an early manifestation of the twentieth century absorption problem. Chilled beef, which became the core of the export trade, had, under proper temperature control, a life of about forty days from plant to table. Since the trip to England normally took three weeks, and part of the remaining period was needed for retail and household handling and storage, around ten days were left to cover loading, unloading, and wholesaling, plus shipping delays. Ships had to be filled quickly, requiring a large daily *frigorífico* output, and meat consignments and ship scheduling had to be nicely timed, with little margin for error. To British and American packers, with their accumulated base of technical and organizational experience on which to build incrementally, it was no doubt a high profit/moderate risk opportunity. To Argentine entrepreneurs and financiers, lacking this base, it was high profit/high risk, and they shied away to more low-risk alternatives, of which there were many.

Did Argentina gain or lose because of this? Had native firms competed in the export trade, net foreign exchange earnings might have been higher, and Argentine organizational, technical, and marketing skills would have been elevated by the need to keep up with the competition. Technological spillover from processing by-products might have been more widely diffused locally. On the other hand, until 1929 Argentina was enjoying an unusually prolonged export bonanza whose benefits, despite important gaps, were spread more widely (if not more equitably) than in the rest of Latin America, apart from Uruguay. The narrowness of its technical and entrepreneurial base was only brought home after 1929, when rapid industrialization was forced on it by the collapse of the export bonanza. Benefits obtainable with a thirty-year delay tend to have a small present value, so hindsight judgment has to be rather hesitant on the issue. Other inactions, e.g., land reform and technical education, merit more certain condemnation.

### *The Chilean Nitrate Industry*

Chilean nitrates are a simpler case. Chileans were active participants in the initial development of the natural nitrate industry of the Atacama desert, then mainly Bolivian and Peruvian territory. Chile declared war on the two, largely at the instigation of Chilean nitrate investors, and wound up annexing the nitrate region from Bolivia and Peru in 1883. But, having instigated an imperialist war, the Chilean *salitreros* proceeded, over the next two decades, to sell most of their holdings to British entrepreneurs.

The denationalizing was not due to prophetic vision by the Chileans. On the contrary, Chile's virtual world monopoly of nitrate production only ended with the appearance of low-cost synthetic nitrates after World War I. In the preceding four decades, nitrates, sparked by expanding world civilian and military demand and by falling freight rates, were an unusually sustained export bonanza.

Neither were there forbidding scale barriers or technological leaps to frighten Chileans into selling out. Throughout the bonanza period, nitrate was processed by 100–150 *oficinas,* averaging fewer than 400 workers per *oficina* and employing the technically simple Shank method of crystallizing nitrate. Refined nitrate required only simple handling, storage, and shipping equipment. The Chileans sold not to giant foreign corporations but to individual Britishers seeking opportunity, although a few of the latter subsequently achieved plutocratic status from nitrates. We are left, therefore, with the uncomfortably simple conclusion that the Chileans had a strong preference for quick capital gains over long-haul profits and acted accordingly when British takeover bids gave them the opportunity to choose.

### *The Chilean Copper Industry*

The third illustration, Chilean copper, though more complex, reinforces this conclusion. Copper was also mainly a locally owned export industry dating back to the early decades of the nineteenth century. Mining and smelting were small scale, exploiting rich, shallow veins by primitive methods. Shallow veins of up to 30 percent copper content enabled Chile to become the leading producer by mid-century, its output averaging over one-third of world production between 1850 and 1880. A few mines adopted imported technical improvements in the course of this expansion, notably the reverberating furnace, coke fuel, and the mixing of sulphide with oxide ores in order to smelt the previously intractable sulphides. But most continued to use the same methods that had shocked Darwin by their back-breaking primitiveness during his Beagle visit in the late 1830s.

High mid-century prices, however, excited a wave of copper prospecting in the United States and Europe and an intensified search for economical methods of refining lower-grade ores. By the 1870s, the output from these new sources helped send copper prices into a decline, and soon afterward, Chilean production. By the 1890s, Chilean production had dropped to half of its 1870s average and to fifth place in international ranking. Yet the long-term trend of copper consumption was firmly upward. The market for copper sheathing had dwindled, along with the fleets of wooden-hulled ships that used it, but electrification, the international arms race, and falling ocean freight rates presaged a bright future for copper.

The basic challenge was to lower production costs by pursuing either of two general alternatives. One was to improve the productivity of existing Chilean mines by incorporating aspects of European mining technology, notably those allowing processing of Chile's rich copper sulphides. Europe in the late nineteenth century was a rich repository of alternative techniques for processing various types of copper compounds and ore compositions on a modest scale. The other alternative was to adopt the recently developed mass techniques for digging, concentrating, and smelting large low-grade uniform ore bodies in which the Americans were the major pioneers.

Hindering the pursuit of the first alternative was the scarcity of technical expertise among most Chilean mine owners and administrators. Few mines employed professionally educated mining engineers, and long dependence on mining had not stimulated serious efforts to train high-level mining technicians; in 1903, enrollment in the nation's three small mining schools totalled only 180. A few Chilean mines modernized, as did some moderate-sized French- and British-owned mining companies, but many of the Chilean-owned mines stagnated or closed shop. Blocking pursuit of the large-scale alternative was the impossibility of getting local finance to assume the heavy capital costs and risks involved.

Chilean landowners with copper on their property looked abroad, engaging agents and promoters to search for foreign investors. They soon learned that able promoters, familiar with both Chilean mining properties and foreign mining capital, were scarce and expected fat commissions for their efforts. Small ventures failed to draw responses. Thus, even though it was technologically possible for Chilean copper to have followed the nineteenth century pattern of incremental improvement and growth, when foreign capital did come in, it was on a twentieth century scale. Two giant mines were brought to fruition by the American Guggenheim interests. One was incorporated into the Kennecott Copper combine, and the other later became the property of the Anaconda Copper Company. Capital-intensive, foreign-owned and operated, they converted Chile into a textbook example of an enclave mineral economy.

Chilean entrepreneurial sluggishness troubled contemporaries. Conservative critics saw it as backsliding; too facile export bonanzas had corrupted the Roman virtues which many attributed to earlier generations of the ruling elite.[6] To radical critics it was merely the normal behavior of landed oligarchs who saw clearly that their economic interests did not depend on the broad economic and technical progress of the economy.

The prolonged export booms of Period I produced their Latin American plutocrats. Some were born with a long lead toward plutocracy; others had it thrust upon them by luck or political skills, but some also qualified as Schumpeterian entrepreneurs. One thinks of Simon Patiño, the Bolivian Indian who made it very big in tin, or of Torcuato Di Tella, the Italian immigrant who created the basis in Period I for what later became the largest indigenously owned engineering goods firm in Latin America. But most Latin American capitalists were molded by the greater security and superior credit status of landed investment, by mistrust of impersonal collective forms of accumulation like the open corporation, and by awed paralysis before the superiority of European and American technological and managerial skills. Prolonged adversity might have forced a modification of these attitudes, as it partially did in the 1930s and 1940s. But during the sustained export bonanzas of Period I, investment behavior guided by the above views brought relatively easy profits and quick capital gains, which, in turn, reinforced belief in the soundness of the views. Contemporary critics were thus on the right track in their perception that the class behavior of Latin American capitalists was hindering more equitable, firmly based economic growth. What still needs to be done to clarify historical understanding is more precise research on the determinants of this behavior.

A final issue is whether it was ignorance or political infeasibility that prevented the adoption of policies to widen the distribution of benefits from Period I growth. On this point one does not have to rely solely on hindsight either. Latin American political leaders of this period were rarely given to detailed philosophizing about development, but there were interesting exceptions whose fate helps illuminate the social forces that cemented economic dualism.

### *Other Development Models*

One of these exceptions was the so-called Generation of 1837 of Argentina. It was composed of leading political figures of the 1860s and

1870s—notably Sarmiento, Alberdi, and Avellaneda who, exiled during the Rosas dictatorship, had participated in its overthrow in 1852. Years of exile had turned them into scholarly social analysts and ideologues of Argentine modernization, and when they rose to high position—two of the three attained the presidency—they attempted to implement their ideas. Their model was the United States, which they regarded, despite slavery and the Civil War, as a socially and economically progressive society. To transform Argentina in this image they proposed as key instruments: massive European immigration; universal primary education; a national network of trade schools and technical institutes; a general reorientation of higher education from law, liberal arts, and theology to pedagogy, science, and engineering; and a broad-based propertied yeomanry to be created by free homestead grants of public lands and by the promotion of European agricultural colonies.[7]

Of the proposals, only massive European immigration, which coincided fully with the rising labor needs of the *estancieros,* was fully implemented. In education, although a network of free primary schools was rapidly extended until, by the end of the century, Argentina's literacy rate was approaching that of Western Europe, the country was not dotted with trade schools and technical institutes as Sarmiento and his *confrères* had urged and the universities continued in the traditional law, medicine, liberal arts pattern until after World War I. A homestead act modelled after that of the United States was passed during the Avellaneda presidency, and some European colonies took root in the 1870s. But, concurrently, the eradication of the plains Indians placed huge expanses of pacified pampa land in the government's hands. It promptly dissipated it by selling it in *estancia* portions at an average price of less than ten cents (U.S.) per acre and by awarding large sections to high ranking veterans of the Indian wars. A rapid, sustained rise of land values soon afterward brought further European colonization to a virtual halt, and the homestead act remained a dead letter. The mixed outcome was closer to the interests of the *estanciero* groups dominating Argentine politics than to the vision of the Generation of 1837.

Also illustrative is the fate of the Balmaceda regime in Chile. In the 1830s, a coalition of astute Chilean conservatives had succeeded in terminating post-independence political disorder by imposing an oligarchic constitution under which presidents with wide powers were elected by a congress, which, in turn, was chosen by an electorate limited to literate property owners. The system gave Chile almost six decades of political stability marred only occasionally by abortive attempts at insurrection. Yet this unique Latin American record, a source of pride to the Chilean aristocracy, who basked in the accolade, the "English of

Latin America," was brought to an abrupt and bloody end when they overthrew President José M. Balmaceda in 1891.

As in most civil wars, the 1891 insurrectionists had a mixture of grievances. The immediate one, on which the congressional majority raised the standard of rebellion, was constitutional: the president was illegally breaking a budgetary impasse with Congress by unilaterally extending the life of the budget beyond its official termination date. Yet in previous administrations, budgetary disputes had always led to compromise solutions which allowed presidential programs to go forward, whereas this time, Congress had become obdurate in denying further funds for Balmaceda's development program. Given the composition of the Congress, most historians are agreed that it was the opposition of wealthy landowners and merchants to the program that was the root cause of Congress' intransigence and of its willingness to turn to armed insurrection.

The major components of Balmaceda's program were: expanded public works, nationalization of the railroads, increased Chilean ownership of the nitrate industry, and establishment of a government-owned central bank. It was the sort of moderately nationalistic developmentalism that had contemporaty counterparts in various countries of the more developed world. But Balmaceda was apparently a Listian whose time had not yet come in Chile. His cause did not awaken mass or even middle-class enthusiasm to counterbalance the hostility of most of the landed and mercantile classes and of the British community. His supporters to the bitter end were mainly aristocratic apostates and some middle-class intellectuals. In the celebrations that followed Balmaceda's resignation and suicide, the popular classes either joined their betters or held their peace. Only with the upsurge of middle-class reformism during the interwar years was Balmaceda belatedly elevated to the status of martyred national hero and were the insurrectionists lowered to the status of unpatriotic villains.

The most successful developmentalists by far were "the best and the brightest" of Porfirio Díaz, the Mexican Positivists who guided socioeconomic policies during the thirty-five years of the Porfiriato era. Alone among Latin American development ideologues, they were able to implement most of their developmental program. Díaz was, of course, of great help. He provided them with brutally enforced law and order, and by re-electing himself successively from 1876 until he was forced to flee in 1911, he gave his economic teams longer political continuity for planning than is the fortune of most *técnicos*. But the successful implementing of Positivist developmentalism and the longevity of the Porfiriato were due, in turn, to the substantial economic gains felt by leading groups who might otherwise have bridled at the corruption and political crudities of the Díaz dictatorship. It was a long, prosperous, progressive era for domestic and foreign landowners, merchants, bankers, mineowners, and industrialists, especially when contrasted with the preceding decades of foreign invasions, defaulted debts, and the inconclusive internal war between conservative and liberal factions that had torn the country apart.

The Positivists (the ideational link to Comte is a loose one) were enthusiastically elitist. They were convinced that the economic and social development of Mexico had to be built on two main foundations. One was the thin stratum of educated Mexicans, whose prosperity had to be promoted by enlarging its ownership of agricultural land, building up the transport and communications infrastructure, improving and expanding secondary and professional education, promoting banks and other financial intermediaries, and keeping property taxes low. The other was European immigration and foreign capital. In contrast to the agricultural bias of immigration policy in other Latin American countries, the Positivist emphasis was on attracting immigrants with industrial skills and capital, partly by means of selective protective tariffs and tax concessions. To attract foreign capital, Mexico's disorderly foreign debt was consolidated and the country's credit worthiness restored. Generous concessions were offered to foreign investors in mining, agriculture, and infrastructure, while the government borrowed heavily abroad to finance its public works projects. For the Positivists, building on the two foundations was necessary in order to minimize the drag on Mexican social and economic progress from the backward Indian and *mestizo* masses. An expanding modern sector, increased exposure to European habits and skills, and the spread of public education would gradually civilize the masses and complete the transformation of Mexico into a modern progressive society. But the Positivists took a long, relaxed view toward the transformation. For example, a free, compulsory primary education law passed in 1891 was implemented at a leisurely pace and chiefly in the large towns and cities. On the eve of the revolution, the Mexican literacy rate was still only about 20 percent.

As a development ideology Mexican Positivism differed fundamentally from that of the Generation of 1837, or of the Balmacedists. The latter two tried in different ways to cut across the grain of the dualistic growth trends inherent in Latin American market forces and distribution of power. The Positivists unabashedly encouraged these trends. In this difference probably lies their far greater measure of success; their policy planning fitted closely the perceptions of those who supplied capital and who mattered politically.

Whatever may be the limitations of the Marxist theory of the state for later eras, it seems to offer the most appropriate summary view of the nature of Latin American governments during Period I. If the Latin American state fell short of being purely the executive committee of the ruling class, it was mainly because regional conflicts of interest and super-charged Latin egoism generated frequent squabbling over the chairmanship.

Calling the behavior Marxian, however, does not get one very far, for the shadings of Marxism vary with the eye of the beholder. Currently, the dominant tendency of Marxist thinking on Period I is to emphasize imperialist pressures and to imply that isolation from contacts with international capitalism would have allowed indigenous forces to carry Latin American countries upward to "true" economic development. In the trenchant view popularized by André Gundar Frank, international capitalism has been the cause of Latin American underdevelopment. Recurrent military intervention in Central America and the Caribbean by the United States, with its penchant for suppressing incipient populist movements by imposing sordid, durable dictatorships on the victimized countries, lends some credence to this view. But for Latin America as a whole these are exceptions rather than the rule. There is little evidence that the views of most of the ruling groups on socioeconomic policy represented anything other than a confident if parochial identification of their own self-interest with that of the nation. Neither is there convincing evidence, at least for Period I, that there was a powerful technological and capital-accumulating potential embedded in Latin American culture. Latin America at mid-century was not on the verge of reinventing the railroad engine or the power loom.

During Period I sustained economic growth required the expansion of foreign markets and substantial inflows of foreign technology and organizational ideas. What gave Latin American growth its excessively dualistic bent was the nature of market relationships between unequals and the unwillingness of Latin American governments to pursue vigorous measures to redress the imbalance. This reluctance must be attributed fundamentally to the parochial self-interest of the Latin American ruling classes. To them the market bargain was not notably unequal. They owned the natural resources, even if foreigners controlled the skills, technology, and finance needed to valorize the assets. To hasten valorization, they learned the need to regularize payments of the foreign public debt, encourage immigration, and create a favorable climate for private investment. To promote the latter, they adopted foreign business institutions and the *laissez-faire* slogans and symbols by which foreign businessmen of the time identified favorable investment climates: free trade, the gold standard, balanced budgets, equal treatment of foreign and domestic business, no unionization, and the like. Reality fell far short of the slogans, for Latin American business groups were as ready as their overseas counterparts to invoke state power in pursuit of their private ends. But the fervor with which they clung to the slogans despite continued backsliding, even after the slogans had begun to lose favor overseas, has charmed at least some observers.[8]

The results paid off handsomely for both parties during Period I. It was the Latin American masses, by-passed by the arrangements, who felt the brunt of the consequences of unequal market relations. While Latin American countries were not equipped at the beginning of the period for broad-based capitalist development, the technological and capital requirements and the demand trends of the nineteenth century were favorable for nurturing the institutional and human resource base for such development. Overseas policy recipes for this development were also available for adoption by the imitative Latin American societies, although they required creative modifications to fit local characteristics. The rejected development ideologies—those of the Sarmientos and the Balmacedas—were attempts to transfer some of these foreign recipes in order to induce more popular-based capitalist development. But these would have imposed short-run costs on the affluent classes and would probably have temporarily damaged the investment climate—sufficient reasons for the ruling classes to oppose them.

There is little basis for believing that imperial threats and pressures played a determining role in the above cases. There is no evidence that the British were responsible for the *enrichez-vous, messieurs* policies of the 1880s, which pushed aside those of the Generation of 1837, although British investors were prominent among the favored *messieurs*. In the Chilean Civil War, the British community helped finance the insurrectionists, but the United States backed Balmaceda diplomatically while the British government remained neutral and British gunboats did not get sailing orders. Rivalry among the big powers even brought support to Mexican liberals in their fight against the French occupation from an unlikely source, their predatory northern neighbor. Without big-power rivalry, the weight of imperialism would, no doubt, have pressed more heavily on Latin American policy making, but such rivalry was a continuing reality during Period I. Neither was international capitalism all that rigidly committed to a single set of behavioral rules for client states. During Period I, a number of small, weak countries outside Latin America, including British imperial dominions, were able to implement reformist measures rejected in Latin America without permanently damaging their credit worthiness and investment climate. In sum, the evidence indicates that at least outside Central America and the West Indies, many progressive policy options were foreclosed by domestic rather than foreign constraints. Technologies were imported, but the political decisions that accentuated their dualistic consequences were homemade.

## Technological Diffusion in Latin America—Period II

Latin America was far more heterogeneous at the onset of Period II (after the 1920s) than at the beginning of Period I. This was mainly brought about by wide differences between countries in the intensity of export growth and in the degree to which they had been able to attract European immigrants during Period I. At the head of the list were Argentina and Uruguay, whose per capita income in the 1920s approached that of Western Europe. They were over 50 percent urbanized, had large middle classes, substantial agricultural processing, clothing, repair, and service industries, and a largely commercialized agriculture. The list descends through Chile, Mexico, Brazil, and Colombia, all of whom had progressively lower per capita incomes and less extensive modern features than the top two, set off additionally by a greater degree of socioeconomic backwardness. It tails off to countries like Peru, Ecuador, and the Central American republics, where primary export bonanzas had done least to complicate the two-class agrarian characteristics with which they began Period I.

The first two groups of countries after the mid-thirties underwent a dramatic reorientation of their development strategy, from promoting primary exports to pushing the production of industrial import substitutes. While enthusiasm for this strategy has since dwindled, and efforts are now under way in all of these countries to modify their industrial strategy (partly by regional integration schemes and partly by encouraging industrial exporting), the payoffs from these alternatives have not been sufficient thus far to induce the countries to dismantle the industrial protection, import controls, and subsidies to foreign exchange-saving industrial investment that have been the policy mainstays of import substitution industrialization (ISI). Moreover, even though the thrill is gone from ISI for the first two groups, it still titillates the tail-end group of countries, where the shift to ISI has been more recent.

Import substitution industrialization came to the two upper groups of countries first for three main reasons. One is that they were the only Latin American countries with large enough domestic markets for industrial imports (as a result of their Period I growth) to attract industrial investors. Severe balance-of-payments difficulties in the 1930s and big power conversion to military production during World War II had reduced the availability of many imports. Second, greater economic growth and larger inflows of European immigrants during Period I had given them a broader accumulation of skills, entrepreneurial experience, and private wealth on which to base an intensified industrialization effort than was the case for the third group of countries. The base was still weak: new industries tended to be high cost, and very heavy protection and other types of subsidy were usually needed to awaken the investor interest. A pool of such investors did exist, however. Until the 1950s, most private ISI investment was undertaken and financed locally, although often with foreign technical assistance and government financial help. The third reason was political. By the 1920s, middle-class nationalist movements and radical trade unionism were complicating the political scene in the first two groups of countries. Initially, ISI was a mainly inadvertent side effect of devaluation and the import controls which virtually all Latin American governments adopted when primary export markets collapsed in the 1930s. The new political groups, however, quickly adopted deliberate ISI as the core of their long-term development strategy. In some of the countries they gained poltical ascendancy in the 1930s and 1940s, but even where they did not, politicians felt impelled gradually to adopt ISI measures, partly because the alternatives were bleak economically, and partly to keep their hold on power.

As articulated by its intellectual proponents, ISI was expected to achieve the following major objectives:

1. Progressively reduce the overall import/GNP ratio, thereby allowing GNP to grow faster than primary exports.
2. Increase the pool of industrial technology and skills by successive domesticating of new import substituting industries.
3. Relieve rural underemployment, thus hastening the mechanization of agriculture.
4. Create a national industrial bourgeoisie, whose self-interest in enlarging the national industrial market would make it a powerful political force for overcoming the opposition of the landed oligarchs to ambitious public sector development programs.

The ISI strategy was expected to unfold in two phases. In the first, the stress would have to be placed on producing consumer goods, since these made up the largest proportion of industrial imports and were also believed to be less capital- and scale-intensive than intermediate and capital goods. The expansion of consumer goods industries would require, initially, imported inputs, which could be paid for with the exchange savings of the established ISI consumer industries. But the expansion would also progressively enlarge the home market for import substitution of intermediate and capital goods. This second phase would be open-ended, since mills would be needed to build more mills ad infinitum. In the 1950s, the Economic Commission for Latin America created development plans for a number of countries around this sort of staging. Although the plans were not followed closely, the rationalization of ISI gave intellectual support to the more opportunistic pursuit of the strategy adopted by the governments.

By the end of the 1950s, it was becoming evident that results were falling far short of expectations. Manufacturing had become the largest and the fastest growing commodity sector in terms of output, but remained behind agriculture and services in employment in the first two groups of countries. More ominously, the rate of growth of industrial employment was also slowing down. Since the anticipated rural labor shortage was not developing, such farm mechanization as was taking place was only adding to the already heavy rural underemployment and was pushing more rural migrants to the cities. There, they were being absorbed mainly in service activities with few skill or capital barriers to entry, in unremunerative jobs, and in self-employment. Socially, they were rapidly enlarging the slum areas, making socioeconomic dualism more visible in the cities.

Much of the problem was on the supply side. The acceleration of demographic growth that had begun in the 1940s in most of Latin America was beginning to feed recruits to the labor force at an increasing rate by the end of the 1950s. But a good part of the problem was the slackening growth of demand for labor by industry and other modern sectors. Open unemployment was rising, as was low-paid "disguised" unemployment, and the "discouraged worker effect" combined with a rising dependency ratio to lower the labor force participation rate.

As for general distributive effects, ISI had created new jobs and skills and raised the incomes of a fair portion of the population. A cautious generalization one could make from graphing ISI efforts would be that the 1920s and 1960s Lorenz curves intersected, usually with some further rise of high Gini inequality coefficients. The intersection in most industrializing countries was probably around the seventieth percentile, the families from near the intersection point to about the ninety-fifth percentile gaining relative to the rest. The trend, however, was probably not monotonic. Gini inequality probably declined between the 1920s and the 1940s, but rose again between the 1940s and the 1960s. The post-war trend was probably toward greater inequality, with the intersection point moving to the right.[9]

Another major disappointment was the leveling off of the import/GNP coefficient by the 1950s in most of the industrializing countries. Growing faster than the export growth rate now required increased foreign direct investment and portfolio borrowing. The external indebtedness of the industrializing Latin American countries began a strong upward trend, and by the mid-1960s the foreign exchange costs of servicing the foreign debt and covering dividend and royalty transfers had become a matter of chronic concern for both borrowers and lenders. In pursuit of economic independence, the industrializing countries had slipped again into the situation where the need to prove credit worthiness to foreign financiers reduced their freedom of action.

That other rejected development recipe of Period I, the creation of a favorable climate for foreign investment, had to be refurbished also, since the progressive national bourgeoisie were proving unequal to the task of carrying ISI into the more capital- and scale-intensive reaches of the industrial gamut. Regulation of foreign investment was again liberalized in the 1950s, and, contrary to the wisdom of the 1940s, foreign corporations proved quite willing to invest in manufacturing. So willing were they, that by the end of the 1960s complaints were widespread among former ISI enthusiasts about the denationalization of industry. Denationalization was indeed occurring, with multinationals pursuing a particularly aggressive take-over policy in tobacco and food processing. Mainly, however, foreigners were dominant or competed with state-owned firms for dominance in newer and more technically sophisticated industries: motor cars, heavy engineering, petrochemicals, synthetic fibers, oil refining, agricultural machinery, and pharmaceuticals. In all, they owned a fairly large share of the industrial sector—as much as one-third, according to some estimates for Brazil and Mexico. In addition, locally owned firms were running up foreign exchange outlays for patent and trademark licenses and technical assistance.

Explanations of these disappointing results may be divided, as may virtually all analyses of Latin America's economic traumas, into two categories: those stressing price distortions and those stressing structural factors. The first approach tends, in my view, to overplay both the capacity of Latin American governments to bring about enduring shifts in relative prices through conventional macro-policy instruments and the speed of adjustment of Latin American firms to changes in relative prices. Parroting "your exchange rate is overvalued" to countries that have been devaluing frequently during a year, using all sorts of imaginative tactics, becomes rather ludicrous.

The key to understanding the difficulties lies in some overlooked dynamics of ISI. One is that ISI, as practiced in Latin America, tends to lead to a high rate of shifting of the consumption mix in an import-intensive direction. The drastic reduction of imports of consumer manufactures does not herald the decline of consumer ISI, but rather its continuation by other means. Experience gradually teaches industrial firms to assess local market opportunities without the help of direct imports. As they sharpen their assessment power and increase their contacts with overseas sources of product design and trademarks, these firms gradually shift their investment in consumer manufacturers from replacing imports to replacing earlier import substitutes. Various chains can be identified: cotton textiles, rayons, nylon, polyesters; tube radios, transistorized sets, black-white television, color television; metal household goods, plastics, etc.

The chains of consumer goods substitution also engender import substitution of intermediates and capital goods. But since these are largely induced by input requirements of the consumer goods producers, they occur with varying lags, whose duration depends on the size of the home market for the backward-linkage goods, the availability of investment finance, and the terms on which foreign suppliers are willing to provide the requisite technologies for local production. New consumer goods tend, therefore, to be initially more import-intensive than goods they replace. Therefore, while backward-linkage ISI tends to lower the import/GNP ratio, the substitutive consumer goods chains push upward against the ratio. Significantly, the leveling off of the ratios in various Latin American countries in the 1950s occurred when the direct imports of consumer manufactures had been reduced to minimal levels, to what might be called the smuggler's floor.

The dynamics are, in large part, the effect of income elasticity set off by rising household income, but they also reflect a continual shifting of consumer preferences toward new foreign-styled goods—the international demonstration effect. Much of this effect is inherent in the one-sided cultural impact of high income technologically advanced societies on backward ones, but it has been magnified in Latin American countries by the naiveté of the ISI strategy they have pursued. Essentially, the strategy has been directed toward closing home markets to competitive imports (but has been open to the inflow of preference influences), has eagerly promoted the importation of technologies and business practices to meet the new demands, and has been blind to the possibility that many of the business practices are oriented to stimulating new preferences. There are undoubtedly rates of demand shifting that are within the digestive capacity of the imitative Latin American economies, but the dynamics of the process have led to chronic overshooting, which takes the form of excess demand for imports and an accumulating foreign indebtedness that forces periodic economic slowdowns.

The chaining also accelerates the trend toward increasing capital intensity, since the succession of newly domesticated goods replicates the capital-intensive trends of the advanced countries that are providing the product designs and the technologies for making them. Excessively rapid demand shifting, therefore, helps slacken the growth of industrial employment relative to output.

Analysis of the chaining dynamics also prompts looking at the industrial markets that have grown up under ISI. These are mainly oligopolistic—price competition is eschewed, and the competitive game is played primarily along product differentiation and goods promotion lines, most intensely in consumer goods industries. Oligopoly is a pervasive fact of life of twentieth century industrial capitalism, and the small markets of Latin American countries impart a particularly strong and probably unavoidable impetus toward oligopolization. However, whereas firms in the advanced capitalist countries obtain their new product designs and technology through creativity and mutual exchange, Latin American firms chiefly import the necessary ingredients for pursuing oligopolistic competition.

Competition under ISI is strongly biased against investing in domestic technological creativity. Foreign industrial subsidiaries are alternative ways for multinational firms to market their overseas products when direct exporting is blocked by ISI protectionism. Although the multinationals have ample ability to redesign products and processes for local markets, there is little incentive for them to undercut returns from the already incurred costs in their centrally developed products and processes. Rather, the incentives are to extend the payoff from these costs by successively drawing on products from their already developed pool for domestication through ISI. The denationalization of Latin American industry has not been due solely to the financial and productive superiority of multinational firms. It has also been rooted in the high profitability of transferring already developed products at low marginal costs to manufacturing subsidiaries in the Latin American countries and in the superior marketing sophistication of the multinationals in promoting local demand for these products. Local firms have to compete in this game on less advantageous terms, since the prices they pay for imported designs and technology are well above their marginal costs to the overseas suppliers. But the designs and the productive processes are tested and predictable, whereas investing in home research and development is not. As long as heavy import protection permits local firms to recoup the high but predictable costs of importing designs and processes, there is little incentive for them to invest in developing their own.

Accumulating evidence supports this perception of the marketing game. Latin American firms still do little spending on research and development, and the little they do is mainly for troubleshooting.[10] Foreign experts characteristically identify the marketing approach of the industrial firms as "skimming the cream" from the higher income segments of their markets. Data on executive salaries indicate that the remuneration for executives in marketing divisions tends to be considerably higher than that for executives in production and designing divisions and that the differentials have been widening over the past decade. The main reason for this seems to be that high-powered teams can be imported to meet major production crises, leaving the local cadres in charge of the more routine operations, whereas marketing requires more assiduous local supervision. Even the occasional visitor to the major Latin American cities cannot help being impressed by the evidence that the range of foreign-designed products has been widening rapidly and that the strident goods promotion practices of advanced countries are permeating the mass media of Latin Ameriça.

A summary overview of Period II in the industrializing countries shows that they have made considerable progress in building up their technological and entrepreneurial capacity for becoming progressive nineteenth century industrial capitalistic dynamos. There is now, in the larger Latin American countries, a strong, though largely latent, capacity to engage in technological adaptation and creation along nineteenth century lines. Some of this is occurring among small firms who lack the means to search out foreign suppliers and enter into profitable licensing arrangements. Some firms worked out ingenious, if often crude, solutions when World War II suddenly opened up many market opportunities, but also cut off access to foreign solutions. Unfortunately, mid-twentieth century technological progress operates from a set of institutional requirements and procedures different from that of the nineteenth century, and industrial market structures, reinforced by ISI, militate against their rapid development.

Concerned with their continued dependence on imported technologies, many Latin American countries have been investing more heavily since the 1950s in university-level scientific and technical training and in national research institutes. Thus far, however, the investment in high-level human capital has augmented the Latin American "brain drain" more than the flow of domestically produced innovation. The gestation period for developing a significant local innovative capability is a long one, and some see the meager results to date as normal features of an unavoidably prolonged apprenticeship stage. However, like other products, technological creativity is determined by demand as well as supply forces. The "normality" of a century-long period, during which major transformations of the productive structures of Latin American countries have been unaccompanied by significant reduction of their dependence on imported technology, is surely suspect. It suggests that beyond investing in the expansion of technical talent, the Latin Americans need to modify the market dynamics that have been retarding the effective use of their talent.

Yet, there is no easy set of policies for altering this enduring dependency pattern and its dualistic consequences. If the Latin American countries persist in their mercantilistic capitalism, broadening technological diffusion would require them to selectively ration access to foreign technology, curb the preference-promoting aspects of product differentiation competition, and modify income inequality in order to build up internal mass markets. At the same time, they would have to use import liberalization and export promotion to pressure firms into greater productive efficiency and creative technological effort—all this while simultaneously sustaining the investment climate and the overall level of economic activity. This tall order would obviously place novel demands on the astuteness of the political leaders and the state bureaucracy, as well as on the flexibility of Latin American political and cultural institutions.

On the other hand, a trend toward socialist solutions would shift the main barriers from the demand to the supply side. Socialism would drastically alter the market structure, curb the product differentiation game, and distribute the benefits of existing technology more equitably. But it would also turn off, at least for a time, foreign channels of new technology and finance and would transfer the main burden of improving managerial efficiency and technological creativity from the private entrepreneur to the government bureaucrat. This would require a rapid and sustained increase in the efficiency and creativity of Latin American bureaucracies, which, to date, have not been distinguished for either quality.

## Notes

1. For a view that Latin American political stability has been both normal and "efficient," see Charles W. Anderson, *Politics and Economic Change in Latin America* (Princeton: Van Nostrand, 1967).

2. Cf. Brinley Thomas, *Migration and Economic Growth* (Cambridge: University Press, 1954).

3. This scenario is developed in greater detail in David Felix, "The Technological Factor in Socioeconomic Dualism: Toward an Economy of Scale Paradigm for Development Theory," *Economic Development and Cultural Change,* 25 Suppl. (January 1977):180–211.

4. Cf. James Tobin, "Inflation and Unemployment," *American Economic Review* 62 (March 1972):1–18.

5. The estimate is from the *Anuario Estadístico* (1920) as quoted in Markos Mamalakis and Clark W. Reynolds, *Essays on the Chilean Economy,* (Homewood, Illinois: Richard C. Irwin, 1965), p. 219.

6. Cf. Francisco Encina, *Nuestra Inferioridad Económica* (Santiago, 1913) and Alberto Edwards, *La Fronda Aristocrática* (Santiago, 1927).

7. The first Argentine population census in 1869 is prefaced by a long, unsigned essay that trenchantly criticizes Argentine educational and other institutions and proposes reform measures along these lines. This unusual introduction to an official volume of population statistics was reputedly written by Sarmiento, then president of the republic.

8. Cf. Albert O. Hirschman, *Journeys Toward Progress* (New York: Twentieth Century Fund, 1963), pp. 263-83.

9. The trends can be inferred from the careful compilation and analysis of available Latin American income distribution estimates in United Nations Economic Commission for Latin America, *La Distribución de Ingreso en América Latina* (Santiago, 1971), although a good deal of conjecturing has to be done to bridge data gaps. For Mexico, see time series estimates in David Felix, "Income Inequality in Mexico," *Current History* 72 (March 1977):111-14, 136.

10. Jorge Katz, *Importación de Tecnología, Aprendizaje Local, e Industrialización Dependiente* (Buenos Aires, 1972).

# 4
# Overcoming Technological Dependence in Latin America

*James H. Street*

Current analysis of the developmental problems of Latin America has drawn attention to the tremendous gap that exists between that region and the regions of North America, Western Europe, the Soviet Union, and Japan in the degree to which modern science and technology have been incorporated into the culture.[1] Most of the elements in the partial industrialization of Latin America and the region's advances in modern medical care, agricultural productivity, and scientific research have been transferred in their original form from other parts of the world. Relatively few genuinely innovative contributions to these fields can be identified as of Latin American origin.

This gulf is not easily explained, since in pre-Columbian times the Andean and Middle American cultures, although they lacked the use of iron and steel, the wheel, and inanimate sources of power, were nonetheless richly inventive. During their maximum growth periods prior to the sixteenth century, they achieved the capacity to maintain dense and stable populations with dependable supplies of staple foodstuffs and other economic requisites, they continuously elaborated on techniques of textile and pottery making and structural design, and they reached significant levels in computation, astronomic observation, and the recording and transmission of data.

Why and under what circumstances did the interest in maintaining and extending an indigenous process of discovery, invention, and application die? The Spanish conquest and its suffocating institutions no doubt played a major role, yet probably do not constitute a complete explanation. Nevertheless, the long quiescence of a native technological interest and the consequent failure to accumulate a storehouse of proliferating artefacts (characteristic of a developing society) were factors in delaying the advent of the Industrial Revolution in Latin America.

In modern times, the region, at least in its coastal periphery, has had continuous contact with European civilization through trade and has experienced a considerable inflow of immigrant population from regions already familiar with the benefits of industrialization. Although many of these contacts were with the less advanced countries of Europe, including Spain and Portugal, they might have been expected to stimulate greater interest in domestic innovation than actually occurred.

This chapter will consider some of the cultural factors that predispose a society to make effective use of borrowed tools, machines, and processes and that eventually enable specialized groups within the society to participate actively in the innovative process by which invention and discovery are advanced. The history of the Japanese people from the period in which they emerged from the comparatively closed society of the Tokugawa shogunate in the last century to today (when they are noted for their adaptability and innovativeness) has often been cited.[2] Originally heavy borrowers of foreign science and technology, the Japanese have become important contributors to the present world stockpile of knowledge. This shift could not have occurred without significant changes in social attitudes, educational institutions, vocational apprenticeship, and other forms of functional instruction, as well as in the accumulation of artefacts, from which further technical recombinations are built up. These historical changes are now well recorded.

In Latin America, this process of acculturation to the requirements of an industrially diversified society has been very uneven and late in reaching fruition. A review of some of the major educational movements and their relation to technological growth may help us understand this retardation and highlight some recent positive initiatives that may permit Latin Americans to recover a greater degree of autonomy over their own technological growth process.

## The Dependency Explanation

In the recent period of concern with the economic development of Latin America, attention was first strongly focused on the technological gap by members of the structuralist school, among them, Raúl Prebisch and Aníbal Pinto.[3] They considered this gap one of the bottlenecks to growth and urged both an improvement in the means of transfer of technology to Latin America and the creation of domestic sources of scientific and applied knowledge to enhance the economic independence of the region. Yet, even with the vigorous support of the Inter-American Development Bank and help from other sources, progress in creating new centers for research and development and for advanced training in technical fields has been painfully slow.

Much of the recent literature on technological transfers has adopted an increasingly pessimistic tone and reflects a tendency to attribute the condition of technological dependency entirely to the domination and exploitation of the region by foreign centers of financial power and by the great multinational corporations which produce and control much of the know-how essential to industrial growth.[4] Indeed, Osvaldo Sunkel places this relationship at the center of his dependency analysis. He provides the following explanation:

> The capitalist system is in the process of being reorganized into a new international industrial system whose main institutional agents are the multinational corporations, increasingly backed by the governments of the developed countries. This is a new structure of domination, sharing a large number of characteristics of the mercantilist system. It tends to concentrate the planning and deployment of natural, human, and capital resources, and the development of science and technology, in the "brain" of the new industrial system, i.e., the technocrats of international corporations, international organizations, and governments of developed countries.[5]

Sunkel believes that there has been a fundamental shift in the means of obtaining foreign inputs. In earlier decades, Latin American countries were able to acquire elements of their own development in piecemeal fashion through such means as immigration, public financing, and licensing, but, in recent times they must buy "complete packages" of entrepreneurship, management skills, design, technology, financing, and marketing organization from foreign corporations. As they struggle to achieve economic independence by import substitution strategy, they merely provide more protection for foreign subsidiaries, who eventually dominate local markets. Meanwhile the multinational corporations centralize research and decision making in the home country. "The 'backwash' effects may outweigh the 'spread' effects, and the technology gap may be perpetuated rather than alleviated."[6]

These are persuasive arguments, but they concentrate exclusively on external factors. Their acceptance may lead to defeatism and misguided national policies within Latin America. These tend to cut the region off from essential growth sources at this stage of development. It need not be assumed that the world storehouse of useful knowledge is so readily barred to the uninitiated. Internal factors affecting the prospects for domesticating the technological process must also be considered.

## The Technological Demonstration Effect

Borrowed technology, when combined with available local resources, has profoundly affected the development of a number of Latin American countries. Sometimes a substantial impact has resulted from a single major innovation. Such innovations have contributed to significant structural changes in national economies as well as to the rise of new economic and political interest groups. Because of their demonstration effects in dramatizing the influence of technology on society, it might be expected that such innovations would also influence changes in education, in job training, and in complementary forms of business enterprise.

Colombia was perhaps the first Latin American country in the era of political independence to experience the effect of major exogenous innovations on its internal growth pattern. In a country where terrain presented great obstacles, the shallow-draught river steamboat developed for use on the Mississippi river system in the United States proved extremely useful.[7] The steamboat was introduced into Colombia as early as 1828, and by the latter half of the century became common as the key means of internal transport. It greatly reduced the time and cost of shipping, particularly on the trunk route of the Magdalena River connecting the highlands with the Caribbean coast. For the first time, Colombia could develop a major farm crop, tobacco, to replace the precious metals as its principal export, and a new class of frontier landholders was born.

Later, the penetration of the frontier was extended by the railroad, which opened up new highland areas to coffee production. Coffee soon overtook tobacco as the main export, but, unlike the pattern in Brazil, it was not produced under plantation conditions, but by an army of newly settled small landowners. The introduction of cheap barbed wire and new varieties of feed grains facilitated commercial cattle growing and thus laid the foundations of a diversified agricultural and commercial economy.

The growth in the use of the steamboat and the steam locomotive had strong ancillary effects in the United States. Since both employed horizontal boilers and similar means of driving gear, improvements in steamboat engines were rapidly applied to locomotives, and the establishment of a major boatbuilding industry at Pittsburgh contributed to the growth of the U.S. iron and steel industry.[8] Improvements in metallurgy were, in turn, applied to making larger and more powerful steam engines and locomotives. The machine tool industry was also stimulated when two watchmakers, Phineas York and Matthias W. Baldwin, applied their skills in making precision gears to the technical problems of larger engines. An entire apprenticeship system grew up around these activities in the northeastern United States and the Mississippi Valley. Civil engineering also expanded as transcontinental railroad routes were laid out and the Mississippi waterway system was brought under control. The U.S. Army Corps of Engineers assumed a major technical role in internal development in dredging channels, constructing locks, and building levees for flood control.

There is little evidence that the introduction of the steamboat and railroad as major innovations in Colombia had similar effects in stimulating domestic technological activity or functional education. William McGreevey ascribes Colombia's failure to enter a sustained growth period on the basis of the new systems of transport and commercial agriculture to a congeries of human error in governmental decisions and to the incessant civil violence engendered by ideological conflict between Liberals and Conservatives.[9] As a consequence, the educational system suffered abrupt and destructive shifts from ecclesiastical to secular control and back again. In the end, technological innovation provided little demonstration effect, except on some local industry in Antioquia and in the development of the port of Barranquilla, and there was scant stimulus to native industrialism or to technical education.

Examining the same question, Frank Safford concluded that a rigid hierarchical social structure and aristocratic social values in Colombia impeded interest in technical studies.[10] Before 1935, Colombian technical schools offered practically no specializations aside from civil engineering, and students wishing to study industrial, chemical, or petroleum engineering had to go abroad.

In the modern industrial era, which did not actually become significant in Colombia until after World War II, virtually the entire complement of equipment and technical know-how had to be acquired from foreign sources, although an effort was made to increase vocational education through a national apprenticeship system (Servicio Nacional de Aprendizaje). Education in specialized technical fields other than civil engineering was not effectively organized by Colombian universities until the late 1940s.[11]

Argentina began to receive the benefits of the Industrial Revolution somewhat later than Colombia, but experienced a much stronger and more diversified growth effect that reached a peak at the time of World War I. Argentina was the first country in Latin America to attempt to mechanize its agriculture. It also stands out in this formative period as the single country in Latin America that appeared to be developing an educational system strong enough and diversified enough to support an emerging industrial economy under domestic direction and control. Unfortunately, this impetus was later to lose much of its developmental force.

The key innovations enabling Argentina to reach an extremely high growth rate from about 1870 until 1914 were the introduction of British breeds of sheep and cattle, the construction of railroads and artificial ports, the use of refrigeration in packing houses and transatlantic steamships, and the introduction of harvesting machinery on the great wheat farms.[12] All of these innovations were of foreign origin and were initially under foreign management.

During the dynamic growth period, the port of Buenos Aires became a throbbing metropolitan center, with steam power plants, an electrified urban transport system (that included an early underground railway), an extensive telegraph network linking the capital with the interior, and the world's longest transoceanic cable, connecting Buenos Aires directly with Europe. Industrial development was nonetheless limited to light industries producing shoes, cloth, soap, and food products. So great was the flow of foreign exchange from agricultural exports that it seemed unnecessary to the *estanciero* class to create indigenous sources of invention and discovery. Everything useful could be bought for sterling and the gold peso, and technical know-how was the business of foreigners.

## Educational Movements Related to Technological Growth

### *The Sarmiento Movement*

A few Argentine leaders recognized the need for modernizing the country, improving the educational system, and amplifying the rudimentary acquaintance of the general population with science. Outstanding among these leaders in the latter part of the nineteenth century was Domingo Faustino Sarmiento, who waged a lifelong campaign to "civilize" Argentines and to establish a free, informative press. When he became president of the republic in 1868, he vigorously promoted a great variety of scientific and cultural activities. During his administration, a national academy of sciences was organized, physics and chemistry laboratories were established, and engineering instruction was introduced into the newly created naval and military academies.[13] The study of natural history was stimulated by the building of museums stocked with exhibits brought from Europe as well as collections of fossils and specimens of domestic plant and animal life.

Sarmiento created a national astronomical observatory in Córdoba and sent for a North American friend, Benjamin Gould, to direct it.[14] During his six-year term in office, nearly one hundred free public libraries were established and provided with books distributed by a national library commission. Such activities on behalf of a better informed general public were virtually unique in the Latin America of the day.

New hospitals and an institute for the deaf were constructed during Sarmiento's fertile administration, and the basis was laid for domestic research in pharmacy and medicine. This ultimately gave Argentina its principal distinction in a field of advanced science—when Dr. Bernardo Houssay claimed a Nobel prize in physiology in 1948.

Above all, President Sarmiento, with the collaboration of his minister of education and successor as president, Nicolas Avellaneda, promoted a widespread system of popular education, which, together with an increasing number of daily newspapers and free libraries, brought Argentines into contact with the outside world and made Buenos Aires a cosmopolitan cultural influence throughout the region.

While serving as his country's diplomatic representative in the United States during the 1860s Sarmiento had been much impressed with the movement led by Horace Mann to establish teacher training schools and to foster universal public education in an expanding frontier population. After his return to Argentina to assume the presidency, Sarmiento undertook to propagate Mann's philosophy and to reproduce the popular education movement in his own developing nation. With the aid of a staff of young women selected in the United States for their familiarity with the public school system, Sarmiento pushed the rapid construction of schools in many parts of the hostile frontier as well as in Buenos Aires.[15] School attendance was made compulsory throughout the republic, and enrollments nearly doubled, although among working-class families and especially those in the interior, the dropout rate was high.[16]

Embued with enthusiasm, the teachers from North America introduced educational methods that were the most advanced of their time; among their educational innovations were evening classes for working adults. Argentines were thus given the opportunity to become a functionally literate people well before their contemporaries throughout Latin America. The national literacy rate rose from 22 percent in the census of 1869 to 65 percent by 1914.[17] However, because the Argentine popular education movement went into decline, and there was a rapid growth of population (increasingly made up of immigrants during this period) from 1,900,000 in 1869 to 7,900,000 in 1914, there were actually twice as many illiterate Argentines, in absolute numbers, in 1914 as in 1869.

Sarmiento's influence was also felt for a time in the neighboring countries of Chile, Uruguay, and Paraguay. President Manuel Montt of Chile invited him to establish a normal school in Santiago that incorporated his progressive outlook. Sarmiento was disappointed, however, in the direction taken by Chilean higher education. The national University of Chile was dominated by the classical outlook of its first rector, the eminent Venezuelan Andrés Bello, who was deeply immersed in juridical studies. Even though frequently drawn into political controversy, the University of Chile remained essentially traditional in curriculum and contributed little to industrial growth.

*The Positivist Movement*

As the nineteenth century advanced, Latin American intellectuals in touch with European currents of thought were excited by the positivist philosophy of the French sociologist Auguste Comte and of other writers who elaborated his view. The resultant interest in scientific method led to the founding of a number of local scientific societies, but the research focus was soon dissipated by the broader interest in moral questions and purely literary expressions. In Brazil, the new positivist "religion of humanity" under the leadership of Benjamin Constant Botelho de Magalhaes became a vehicle for republican political sentiment and, in economic terms, for a shift to a southern commercial regime based on coffee and cattle rather than industrialization.

In Mexico, the positivist movement was associated with the vigorous expansion of railroads, mining, and petroleum development under the regime of Porfirio Díaz. The leading *científicos,* led by José Yves Limantour and Justo Sierra, enthusiastically sponsored an elite interest in scientific education, yet neither they nor President Díaz saw the indigenous and mestizo populations of Mexico as having any but a servile role in the development of the country. The new industries were necessarily directed by foreigners, and mass education was obliged to await the revolution.

*The Rébsamen Movement*

A noteworthy exception to the general lack of interest in popular and scientific education was the movement begun by the Swiss educator Enrique C. Rébsamen at Xalapa in the state of Veracruz in the 1880s.[18] Rébsamen, a disciple of the early reformer Johann Heinrich Pestalozzi, founded a normal school that was to have an enduring influence in the region, although it was discredited for a considerable period by leaders of the revolution as a foreign, bourgeois activity. A precursor of John

Dewey's "learn-by-doing" method, Rébsamen thought of education as a functional process in which pupils should be exposed directly to the materials, plants, and animals of their natural environment from their earliest years. Under his method, they were taught to appreciate these elements as features of natural history, as sources of esthetic satisfaction, and as materials for practical use. Thus prepared, the child was expected to make better functional use of the resources at hand in his own community.

Rébsamen attached to his clssrooms an array of shops and laboratories in which children worked with their hands in the arts, crafts, and sciences. Additions to rural schools were built by the children themselves under the guidance of their teacher-craftsmen, and each school was surrounded by gardens, orchards, and livestock pastures as a self-sufficient enterprise.

Interest in the Rébsamen method as related to community development has recently been revived and is making considerable impact in Mexican rural education. So isolated are some village schools that teachers must often begin their instruction in an indigenous language, such as Nahuatl or Totonaco. Unfortunately, the number of well-prepared secondary teachers outside the major cities of Mexico has so far been insufficient to permit functional education much beyond the handicraft stage of instruction.

## The Decline in Technologically Oriented Education

The clarity of Sarmiento's vision of what was necessary for Argentina's scientific and technological growth was gradually obscured as the country became wealthy in the years before World War I. Educational methods at the primary and secondary levels became routinized, and, except for a few well-supported preparatory academies, public education suffered considerable neglect.[19] Teachers were rarely employed full time and had to supplement their incomes with other jobs. Vocational and agricultural education at the secondary level was not introduced before the late 1920s. Class sessions in the elementary grades were held only in the morning or the afternoon (and half-day sessions are still the practice today, except in privately supported preparatory schools).

As late as 1931, despite regulations calling for mandatory attendance, 25 percent of the children of elementary school age were not attending school.[20] Attrition rates throughout the primary, secondary, and higher levels of education in Argentina continued to be extremely high in comparison with countries such as France and the United States. In

1959, of 1,000 Argentine students who had begun the first grade, only 261 were still in school in the seventh year, as compared with 420 in France and 800 in the United States.[21] The attrition rates for other Latin American countries were generally higher, above all in the rural areas.[22]

Despite the introduction of popular education, the Argentine universities retained an elitist and traditional character that, until the Reform of 1918, stultified most efforts to promote scientific research. Professorships were often awarded for purely honorary reasons, salaries were nominal payments for part-time work, and the full-time investigator was rare. In a pattern that was to become familiar in the universities throughout Latin America, enrollments were heaviest in the Argentine faculties devoted to philosophy and letters and to the professions associated with positions of social status: law and medicine, and to a lesser degree, accounting and architecture. Agronomy, animal husbandry, and viniculture were neglected, while the newly founded scienfitic institutes were undersupported. Nothing resembling the land grant college movement in the United States, with its emphasis on the agricultural and mechanical arts, was undertaken in Argentine higher education during the development period—nor, for that matter, elsewhere in Latin America.

Given the Argentine orientation toward Europe and the growing wealth of the *estanciero* class, it was understandable that when the children of the new aristocracy were sent abroad to study, they were able to attend the most prestigious institutions of the Continent and the British Isles, but chiefly for cultural finishing. Sunkel has described this form of education—common among Latin American upper-class families—as essentially "ornamental."[23] It had little functional relation to the developmental needs of growing frontier nations. Rarely was foreign education directed toward preparing a generation for becoming industrial managers, scientific investigators, or the technicians of a new society. This commentary should not be taken as a reflection on the native intelligence or capacities of the Argentines who studied abroad; by background and by personal aspiration, their interests simply lay in other directions.

Alejandro Bunge, who carried on an extended journalistic campaign for the industrialization of Argentina before and after World War I, deplored the lack of attention to scientific, technical, and vocational education during this period. "Argentine education," he complained, "through the end that it sought—the mere diploma—presumed a country already formed. Yet far short of that, barren and poor, the country became populated by lawyers."[24]

When students at the highly traditional University of Córdoba rebelled and put into effect the University Reform of 1918, they were presenting a strong reaction to educational elitism and its accompanying privileges of social advancement.[25] The reform spread with amazing rapidity to other universities in Argentina and throughout Latin America (except in Venezuela, where it was forcibly put down). It had a number of beneficial effects; yet it left unchanged the dominance of traditional fields of study while managing to plant seeds of academic insecurity and administrative instability that endured to the present.

On the positive side, the elitist character of the University of Córdoba was reduced by a more open admissions policy, and aloof, hidebound professors of the older generation were replaced by younger, more flexible ones with a dedication to teaching. The existing curricula were modernized and given a humanistic flavor, and some new subjects were added. Student enrollments, however, remained concentrated in law and medicine.

In an effort to democratize the university, tripartite councils were established representing the teaching faculty, alumni, and students. Rectors and deans became subject to election by these groups, and the autonomy of the university from government intervention was declared. Candidates for professorships were required to present themselves for public lectures, at which their credentials were reviewed and subjected to challenge by the general audience. While this arrangement was designed to insure the relevance of instruction to student needs and interests, it lent itself to popularity contests on the part of aspiring professors and to political manipulation from outside. Perhaps the worst deficiencies of the reform were that it failed to provide an effective evaluation of professional competence by recognized scholars in the field and that if offered no security of tenure, which would have allowed the investigator to devote himself to research and teaching as a full-time career. The effect was to place little premium on the university function of original research.

Alfredo Palacios, the influential young Socialist leader who was later to become rector of the University of La Plata, insisted that the University Reform be given a strong nationalist cast, since it was intended to sever the influence of Europe on Argentine education.[26] In so doing, the leaders of the reform deflected Argentine higher learning even further from the currents of science and technology available mainly from abroad.

For many years the Argentine universities remained centers of political activity, and, particularly during times of crisis, they supplied an outlet for the frustrations of young people, most of whom were already working for a living and thus doubly aware of the retarded development of their own society. The temptation of the government to put down student unrest was strong, and from the time of the massive intervention by the government of Pedro Ramírez in 1943, the Argentine universities were regularly subjected to political housecleaning with every major change of government. Gustavo Martínez Zuviría, the minister of education under General Ramírez whose writings revealed a strong Italian fascist mentality, sought to establish thought control throughout the entire educational system.[27] He restored religious instruction to the public schools for the first time since 1884 and ordered teachers to indoctrinate their pupils with nationalist propaganda.

The trend toward politicizing education was continued under Juan Domingo Perón, who adopted a strongly anti-intellectual line in his first presidency from 1946 to 1955. Despite vigorous student opposition, the universities were occupied, the most eminent professors were dismissed and some of them forced into exile, and their places were given to official polemicists. Requirements for entrance to the universities were virtually eliminated, and a system of monthly oral examinations was introduced that reduced the standards of instruction to those of a diploma mill.

The official disdain for and frequent expulsion of skilled technical manpower extended beyond the universities to other centers of investigation. The Pergamino agricultural research station, one of the few in Argentina, had fourteen professionals in 1949 but lost half of them shortly afterward. Antonio Marino, a scientific plant breeder, had developed two promising hybrid varieties of corn of great potential importance for Argentine agriculture, but his efforts were frustrated because of his personal political views.[28] Raúl Prebisch, one of the world's distinguished economists, was forced to leave Argentina during this period. Such examples could be multiplied. The real loss to the country in potential scientific achievement and practical applications during this turbulent period cannot be calculated.

After the overthrow of Perón in 1955, the government again intervened in the universities, but gradually they recovered their autonomy and updated their curricula. In July 1966, however, a military government headed by President Juan Carlos Onganía again took over the universities. Four rectors of the eight national universities were obliged to resign, and hundreds of teachers were deprived of their positions throughout the university system.[29]

Another upheaval took place under President Héctor J. Cámpora in May 1973 and was continued after President Perón assumed office for the second time a few months later. In the name of popular education, entrance examinations were abolished and the doors of the universities thrown open. Within a year, the student enrollment in the University of

Buenos Aires had doubled, and the national university enrollment had increased from about 300,000 to 450,000 students.[30] Once more, professors of genuine professional distinction, especially those trained in foreign universities, were removed or forbidden to enter the classroom. In the Faculty of Economic Sciences of the University of Buenos Aires alone, 14 ranking professors were dismissed, while some 300 new teachers were hired—few with academic credentials—to staff the crowded classrooms. Within a year and a half, the University was headed by five state-appointed rectors ranging in political identification from the extreme left to the extreme right of the broad Peronist spectrum.

After the death of President Perón in 1974 and the subsequent military take-over, President Jorge Rafael Videla promised to restore autonomy to the universities following one more housecleaning, but political control persisted and was extended to scientific institutes outside the universities, such as the Fundación Bariloche.[31]

Political intervention in the universities was not confined to Argentina. The government intervened heavily in Brazilian universities after the military coup of 1964, and, although they have recently been recipients of increased governmental support, especially in technical fields, they have continued under firm political control.[32] Uruguay's single university was totally shut down in October 1973 after a period of violence that disrupted its normal functioning and was only gradually permitted to reopen. The universities of Chile, noted as centers of free enquiry in Latin America, fell under complete military control in 1973 under the government of General Augosto Pinochet.[33]

The effect of such interventions on professional morale was invariably devastating. Sustained research could rarely be carried on in an atmosphere of political repression, and many investigators had to turn to the relative tranquility of specialized institutes outside the universities to continue their work. The Instituto Torcuato Di Tella in Argentina, the Fundação Getulio Vargas in Brazil, and El Colegio de México fulfilled such functions in the particularly vulnerable fields of the social sciences.

Even where the universities in Latin America have not been politically controlled, they have generally been so poorly supported financially that the choice of an academic career has entailed considerable sacrifice. The battle for academic freedom—in the fullest sense of freedom and support to carry on serious research, as distinguished from simple freedom of political expression—still has to be won in the great majority of Latin American universities and colleges. Even the restricted advantage of strong support for research considered useful to the state, as in the Soviet Union, has been denied to most Latin American investigators. One of the results has been, at times, a "brain drain" of alarming proportions.[34]

## The Consequences of Educational Neglect

In this chapter it has been possible to consider only the more formal modes of education that foster a familiarity with functional approaches for the solution of problems and, hence, a predisposition to utilize technology. A fuller account would describe the informal social conditioning that Thorstein Veblen stressed as necessary for the formation of a "workmanlike" labor force and an inventive culture. Such conditioning includes the play activities of children as well as Veblen's exposure to the "discipline of the machine," or in-plant training and apprenticeship. These activities have been notably deficient in Latin America except in exceptional cases. So have the incentives for invention and discovery by individuals.[35]

The failure to generate a domestic capacity for technological innovation in the general society is often reflected in the practices of domestic industry. David Felix has pointed out that, among the less developed countries, Argentina has the highest ratio of industrial value added to the gross domestic product (GDP), with a uniquely rich endowment of literate, skilled labor and technical and scientific personnel. Yet "most of the industrial equipment used by large firms is of foreign design (although much of the equipment is now locally produced). Large industrial firms, whether locally or foreign-owned, do virtually no research and development beyond that related to troubleshooting adjustments of equipment, materials, and products of foreign design."[36]

Brazilian industrialists likewise conducted little research and development and demonstrated almost no interest in the technical training of their manpower pool until mid-World War II, when, according to Warren Dean, their attitudes began to show a marked change.[37] It has only been recently that they have developed a highly successful apprenticeship system, Serviço Nacional da Aprendizagem Industrial (SENAI). A similar system in Colombia is known by its Spanish acronym, SENA.

## Positive Efforts to Overcome the Gap

In a brief survey, one cannot describe all of the positive actions to improve functional education in Latin America in recent years. Yet three movements stand out. One is the increased attention given to engineering education (beyond civil engineering) in Mexico and Colombia. The Monterrey Institute of Technology and the National Politechnic Institute in Mexico City have expanded their programs and now offer a variety of engineering degrees. The University of the Andes in Bogotá, which opened its engineering school in 1949, has developed a long-term exchange arrangement with several North American engineering colleges. Other Colombian universities have added technical studies to their traditional curricula, and new industrial universities were created in Bucaramanga in 1948 and in Pereira in 1960. Strong programs in the basic sciences have been developed at the Universidad Javeriana in Bogotá under the sponsorship of the Colombian Fund for Scientific Research known as COLCIENCIAS.

The Brazilian government came to the realization in the late 1960s that the nation needed more trained engineers to carry out its ambitious development program. After several years of uncertain action, the government decided to strengthen the university system and establish "centers of excellence" to finance higher education and research by means of a coordinating agency, Coordenação do Aperfeiçoamento de Pessoal de Nivel Superior (CAPES). Under the sponsorship of CAPES and the Brazilian National Research Council (CNPq), an extensive scholarship program was created. During the academic year 1975–1976, 1,200 fellowships were offered in the field of nuclear engineering alone.[38]

A second important development is the increased interest in establishing programs of graduate and upper-level professional instruction in Latin American universities. Few graduate programs were previously available anywhere in the region. Brazil's CAPES plan is a significant effort to remedy this deficiency. Even earlier, graduate training in administrative fields was being stressed in the Faculty of Economic Sciences of the University of Buenos Aires, at the Latin American Faculty of Social Sciences (FLACSO) in Santiago, Chile, and in El Colegio de México in Mexico City. The University of Buenos Aires and FLACSO have unfortunately been set back by recent political developments in Argentina and Chile.

A third manifestation of progress in education is the genuine growth in mass education at the primary and secondary levels in Mexico (particularly in the rural areas) and in Central America. Unspectacular but steady gains in popular education have been made, especially in Costa Rica, Guatemala, and Panama, as a result of the impetus imparted by the movement toward a common market and the short-lived Alliance for Progress. In Guatemala, for example, where 65 percent of the school-age children had never attended school, the number of classrooms constructed in the first five years of the Alliance exceeded the number built in the four centuries since the Conquest.[39]

Classrooms represent only the physical aspect of education; more important is the substantial increase in the numbers of teachers trained, children enrolled, and textbooks disseminated. The commitment to popular education in Latin America fluctuates with national administrations, which control and fund the ministries of education. The present era is a particularly acute one because the population explosion has increased the relative proportion of school-age children in the population and placed an exceptional burden on educational facilities, particularly at the elementary level. Without a wide base of common literacy, effective political participation and productive employment within the younger generation is difficult, and the entire development process is impeded.

Technological backwardness in Latin America is not a condition imposed on the region by recent outside forces, as members of the Dependency School allege. It is a condition deeply embedded in the historical evolution of the culture and fostered by archaic institutions and attitudes; yet it need not continue. Only by persistent efforts to raise the functional effectiveness of education at all levels and thus create the requisite human resources can Latin America overcome its technological dependence and direct the course of its own development.

## Notes

1. The literature on the technological gap is extensive. See for example A. K. Cairncross, *Factors in Economic Development* (London: George Allen & Unwin, 1962), Chap. 11; Julio Broner, "Strategies for Economic Development," *The Process of Industrialization in Latin America* (Guatemala: Inter-American Development Bank, 1969) pp. 258–71; and David Felix, "Technological Dualism in Late Industrializers: On Theory, History, and Policy," *Journal of Economic History* 24 (March 1974):194–238.

2. Asim Sen, "The Role of Technological Change in Economic Development: The Lessons of Japan for Presently Developing Countries" (Ph.D. dissertation, Rutgers University, 1979).

3. Raúl Prebisch, *Towards a Dynamic Development Policy for Latin America* (New York: United Nations, 1963), pp. 11–12; Aníbal Pinto Santa Cruz, "Concentración del progreso técnico y sus frutos en el desarrollo latinoamericano," *El Trimestre Económico* (Mexico) 32 (January–March 1965):3–69.

4. Jorge M. Katz, *Importación de Tecnología, Aprendizaje Local e Industrialización Dependiente* (Buenos Aires: Centro de Investigación Económica, Instituto Torcuato di Tella, 1972); Francisco C. Sercovich, "Dependencia tecnológica en la industria argentina," *Desarrollo Económico* (Buenos Aires) 14 (April–June 1974):33–67.
5. Osvaldo Sunkel, "The Pattern of Latin American Dependence," in *Latin America in the International Economy,* ed. Victor L. Urquidi and Rosemary Thorp (London: Macmillan, 1973), p. 15.
6. Ibid., p. 20.
7. William Paul McGreevey, *An Economic History of Colombia, 1845–1930* (Cambridge, England: University Press, 1971), pp. 41–42, 118–19, 138, 253–57.
8. John W. Oliver, *History of American Technology* (New York: Ronald Press, 1956), pp. 181–95.
9. McGreevey, *An Economic History,* pp. 146-81.
10. Frank Safford, *The Ideal of the Practical: Colombia's Struggle to Form a Technical Elite* (Austin: University of Texas Press, 1976), xiii, pp. 227–42.
11. Laurence Gale, *Education and Development in Latin America* (New York: Frederick A. Praeger, 1969), pp. 53–57, 67–70.
12. Simon G. Hanson, *Argentine Meat and the British Market* (Stanford, Calif.: Stanford University Press, 1938), pp. 11–16, 18–47, 100–101, 117–18; James R. Scobie, *Revolution on the Pampas* (Austin: University of Texas Press, 1964), pp. 77–81. See also David Felix, Chapter 3 in this volume.
13. Hubert Herring, *A History of Latin America,* 2nd ed. (New York: Alfred A. Knopf, 1961), pp. 650–55.
14. Allison Williams Bunkley, *The Life of Sarmiento* (Princeton: Princeton University Press, 1952), p. 468.
15. Alice Houston Luiggi, *65 Valiants* (Gainesville: University of Florida Press, 1965).
16. Herring, *History of Latin America,* p. 654.
17. Arthur P. Whitaker, *Argentina* (Englewood Cliffs, N.J.: Prentice-Hall, 1964), p. 41.
18. Juan Zilli, *Historia de la Escuela Normal Veracruzana* (Tacubaya, Mexico: Editorial Citlatepetl, 1961), pp. 9–82; Edwin Zollinger, *Heinrich Rébsamen, Der Erneuerer der Mexikanischen Volksschule* (Frauenfeldt, Germany: Verlag Huber, 1929).
19. James R. Scobie, *Buenos Aires: Plaza to Suburb, 1870–1910* (New York: Oxford University Press, 1974), pp. 220–25.
20. Henry Lester Smith and Harold Littell, *Education in Latin America* (New York: American Book Co., 1934), pp. 19-32.
21. Robert R. Rehder, *Latin American Management: Development*

*and Performance* (Reading, Mass.: Addison Wesley, 1968), pp. 8-10.

22. John Sloan, "Precedent and Education in Latin America: The Rural-Urban Imbalance," *Educational Innovations in Latin America,* ed. Richard L. Cummings and Donald A. Lemke. (Metuchen, N.J.: Scarecrow Press, 1973), pp. 24-42.

23. Osvaldo Sunkel, "Desarrollo Económico," mimeographed notes for a course in economic development (Santiago, Chile: n.d.), p. 27.

24. Alejandro E. Bunge, *Las industrias del Norte* (Buenos Aires, 1922), p. 173. The translation is my own.

25. The text of the original Córdoba Manifesto may be found in *University Reform in Latin America: Analyses and Documents* (published by the International Student Conference) (Leiden: COSEIC, 1960), pp. 8-12.

26. Alfredo L. Palacios, "A Message to Latin American Youth," *University Reform in Latin America: Analyses and Documents,* pp. 13-16.

27. Herring, *History of Latin America,* pp. 678-79.

28. Carlos F. Díaz Alejandro, *Essays on the Economic History of the Argentine Republic* (New Haven: Yale University Press, 1970), p. 190.

29. John P. Harrison, Joseph F. Bunnett, and George R. Waggoner, *A Report to the American Academic Community on the Present Argentine University Situation* (Austin, Texas: Latin American Studies Association, 1967).

30. See accounts by Joseph Novitski in the *Washington Post,* 28 August and 19 September 1974; and by Jonathan Kandell in the *New York Times,* 19 and 25 September 1974.

31. "El Gobierno Argentino Suprimirá 95 Carreras Universitarias," *Excelsior* (Mexico, D.F.) 28 November 1976, p. 2-A; Nicholas Wade, "Repression in Argentina: Scientists Caught Up in Tide of Terror," *Science* 194 (24 December 1976):1397-99; Wade, "Academy to Campaign Publicly for Oppressed Scientists," *Science* 196 (13 May 1977):741-43.

32. Jeremy J. Stone, "Brazilian Scientists and Students Resist Repression," *Federation of American Scientists Interest Report* 30 (November 1977):1-8.

33. Richard Fagen and Patricia Fagen, "The University Situation in Chile," *Latin American Studies Association Newsletter,* 5 (September 1974):30-37.

34. Morris A. Horowitz, *La Emigración de Profesionales y Técnicos Argentinos* (Buenos Aires: Editorial del Instituto Torcuato Di Tella, 1962).

35. The long-standing failure of Argentine government and industry to encourage domestic inventiveness and creativity through patents or other forms of reward is commented on in James R. Scobie, *Buenos*

*Aires: Plaza to Suburb, 1870-1910.* p. 248.

36. David Felix, "Technological Dualism in Late Industrializers," p. 223. The findings are taken from a detailed study of 200 firms by Jorge M. Katz, *Importación de Tecnología, Aprendizaje Local e Industrialización Dependiente.*

37. Warren Dean, *The Industrialization of São Paulo 1889–1945* (Austin, Texas: The University of Texas Press, 1969), pp. 176–77.

38. James H. Street and Anne Carpenter, *Report on a Survey of the Fulbright-Hays Program in Latin America,* mimeographed (Washington: Council for International Exchange of Scholars, 1975), pp. 2-6.

39. *Perfiles de progreso en Centroamerica y Panamá* (Guatemala: U.S. Agency for International Development, 1965), pp. 26–27.

# 5
# The Economic Case for More Indigenous Scientific and Technological Research and Development in Less Developed Countries

*Dilmus D. James*

The professional economist has long been aware of the impact of technological change. Indeed, this concern with the ex post effects of technological advance has become an orthodox branch of economics. During the last decade and a half, however, economists have evinced increasing interest in the ex ante aspects of technology. This regard for the way knowledge is produced is not surprising. One suspects that the economist has become subject to a powerful gravitational effect, since scientific research and development expenditures have shown a strong growth trend resistant to cyclical changes in the general level of economic activity. In addition, the economist has "backed" into the field of knowledge production via macro-studies of growth by such writers as Abramovitz, Kendrick, Solow, and Denison.[1] These studies have attempted to account for the growth in output resulting from changes in conventional inputs. The large "residual," or growth that could not be explained by the increase of homogeneous inputs, was largely attributed to technological advance. These inquiries naturally led to the exploration of the forces that shape technological change.

Although there are notable exceptions,[2] most literature dealing with science and technology as an instrument for economic development of less developed countries (LDCs) dates from the late 1960s, as a recently compiled "Bibliography on Science and Technology Policy in Latin America" attests.[3] The majority of the approximately 550 entries date from 1969. Most of the scholarly and administrative attention in the field, however, has been directed to the study of technological transfer to the Third World, to the neglect of indigenous technology. This chapter attempts to supplement the sparse complement of literature concerning the economics of acquiring knowledge by internal research and development activities in Third World nations.[4] It will deal with various economic aspects that can provide a rough guide to LDCs in finding an appropriate mix between attempts to acquire scientific and technical knowledge from abroad and attempts to produce it from within. The definition of indigenous technology applied here is a pragmatic one. At the two ends of the poles, there can be little confusion between transferred and indigenous knowledge. If the research is financed and staffed by wealthy nations, and housed exclusively in more developed countries (MDCs), any knowledge produced is a candidate for a transfer; if research is financed and staffed by nationals, and housed in LDCs, it is indigenous.

Most situations, of course, do not fit these pristine categories. The basic research, for instance, may be acquired from MDCs and the basic or applied, developmental, and commercial research may take place in LDCs. A research effort may be partly financed and staffed by foreign aid agencies, international organizations, foreign nonprofit nongovernmental organizations, or foreign private enterprise and combined with financing and staffing indigenous to LDCs. For instance, Marcel Roche reports that thirty-six of eighty-nine employees, or 40 percent, of the scientific and technical personnel of the Instituto Venezolano de Investigaciones Científicas of Caracas were foreign in 1960.[5] In the late 1960s, another writer found that 17.5 percent of all research personnel in Venezuela were foreign.[6]

The phrase "more indigenous research," then really refers to a shift of resources along a whole spectrum of possibilities including acquiring the knowledge in a less developed condition and performing the later stages of research and increasing the financing and staffing of domestic research by LDCs. We could come to some broad agreement as to the degree of indigenous content of each of these dimensions, weight them in terms of importance, and classify various knowledge acquisition projects in some array according to the weighted index of indigenous input. So we are not speaking necessarily of discontinuing transfer of a specific project and performing it in LDCs. In many instances, the move to more indigenous research and development could take the more subtle form of a change in the transfer-indigenous mix at the intraproject margin.

Some comparative data concerning expenditures on research should help us develop a perspective. The Club of Rome, as cited by the Society for International Development, found that over 90 percent of technologists and scientists are working in industrialized countries and that over 90 percent of their activities center on research for the advanced world and findings that are "protected" knowledge.[7] Hans Singer calculated that a little more than 3 percent of the world's expenditure on technical research and development is spent on projects useful to labor-abundant LDCs.[8] Also, whereas MDCs typically spend between 2 and 3 percent of their gross national product on research and development,[9] the figure for LDCs is closer to 0.2 percent.[10] A UNESCO study cited by Ignacy Sachs surveyed twenty-five LDCs and found per capita research and development expenditures at $2.80 in Argentina, the highest for the twenty-five developing countries, with an average of forty-five cents for the entire sample. In comparison, the five most developed Western industrial countries spent about sixty dollars per capita.[11] Another study reported that indigenous research activity in India touched a bare one-eighth of 1 percent of the entire output of the country and estimated that the research expenditure of Indian industry does not exceed 0.1 percent of its sales turnover.[12] Supplemental data pertaining to Latin America is presented in Chapter 2 of this volume.

This chapter will not argue with the generally accepted proposition that most of the Third World's effort should be heavily directed toward applied research and the importation of scientific and technical knowledge. Regarding the latter, I cannot emphasize too strongly that my position is *not* antitransfer per se. Due to specialization, technological transfers move along the lines of comparative advantage among all countries, regardless of the level of development. In a vast number of instances, transferring knowledge into LDCs is the sensible thing to do. Besides, there are some aspects of increased indigenous research that stimulate a complementary increase in the effectiveness of the transfer process. Regardless, the argument in this chapter will concentrate on the absolute and relative emphasis placed on indigenous research by LDCs. It is my contention that the resources devoted to indigenous research and development in LDCs are clearly below optimum in both absolute and relative terms.

Two limitations to this study should be mentioned. It heavily emphasizes Latin American literature, data, case studies, and illustrative situations. This can be attributed partly to the substantial amount of work that has been done on Latin American science and technology policy, but it also reflects the author's own area of specialization. Despite the disproportionate attention to one Third World region, the conclusions are intended to have general relevance for the entire Third World.

Neither does this study attempt to cover the institutional infrastructure needed to support an indigenous research effort. This subject is considered in *Science and Technology Policy Implementation in Less-Developed Countries,* by Francisco Sagasti and Alberto Araoz, which deals with linkages among government agencies, the private sector, educational institutions, and the international arena and their interplay in forming science and technology policy.[13]

**The Analytical Framework**

What does economic analysis tell us of the volume of knowledge that *should* be acquired? Suppose that the application of the new knowledge to future output is used as the criterion for evaluating "bundles," or packages, of knowledge. The maximization principle then becomes that of acquiring and applying bundles of new knowledge as long as the internal rate of return per dollar of cost is higher than the rate of interest, or—more apropos for LDCs—the "shadow" rate of interest is adjusted to measure the internal rate of return on alternative investment projects, such as those in public health or other infrastructure projects.

The internal rate of return for each bundle of knowledge can be calculated if the following are known or can be reasonably estimated: (1) the cost of acquisition and application, (2) the success-failure probabilities, (3) the future stream of additional output if the adaptation and application are successful, and (4) the duration of the payoff period. Both indigenous and transferred knowledge are eventually "capital embodied" (engineered into capital equipment), "human embodied" (transferred into knowledge, skills, and abilities), or "disembodied" (stored in some receptacle for later use, such as books, training manuals, films, computer tapes, and trade journals).

The calculation of the internal rate of return and its comparison with the rate of interest are similar to the exercise conducted for any economic investment. Naturally, there are important differences when dealing with the acquisition of knowledge, not the least of which is Kenneth Arrow's "fundamental paradox of knowledge."[14] When buying knowledge, one cannot know all that the deal entails until the knowledge is actually obtained. In fact, owners of proprietary knowledge go to great lengths to preserve secrecy; in contrast, sellers of capital equipment go to equal extremes to provide information to prospective purchasers. Also, the degree of guesswork increases as one recedes from projects on the threshold of commercial exploitation to fundamental or basic research. In short, some reasonable estimates can be made about the internal rate of return on projects in near readiness to commercial exploitation, but to calculate such prospective social returns for pure research rests on guesses, intuition, and, in some instances, pure faith.

We can visualize the demand curve for scientific and technological research, then, as a definite line and a solid analytical tool at the commercial end of the spectrum. It becomes hazier and fainter as one moves toward the more fundamental phases of research. It is gradually transformed into a broad irregular band with umbra and penumbra, as though it were painted by an artist of the Impressionist School. Whether based on impression or solid analysis, however, decisions are made concerning the amounts of expenditure on various types of research projects. Thus, the following discussion will employ demand curves for useful knowledge as heuristic devices that are roughly indicative of the economic decisions involved.[15]

Figure 1 represents a simplified situation as a point of departure. The panel on the left shows the demand for indigenous research within LDCs; the panel on the right represents the foreign technology available from MDCs. The amount of expenditure on technological transfers is measured rightward from the origin (0), the amount spent on indigenous research is measured leftward. The vertical axis measures the rate of interest (r) and the internal rate of return ($\pi$) on the acquisition and application of bundles of knowledge.

The DD and dd curves in the two panels reflect the expected internal rates of return on bundles of knowledge for indigenous technology (panel on left) and technology that can be acquired through transfer (panel on right). They reflect the demand curves for indigenous and transferred scientific and technological knowledge.

The demand curves for indigenous research slope steeply downward because the internal rate of return is influenced by (a) increasing cost conditions and (b) diminishing returns to knowledge in application. The shallow segment of the $dd_3$ curve of knowledge in MDCs assumes (a) knowledge available in the existing backlog is nonproprietary, or free for the taking, (b) the demand for knowledge has no influence on its price, since knowledge can be dispensed without decreasing its supply, (c) the backlog is not infinite, but its composition is without serious gaps in terms of what LDCs need, and (d) the knowledge can be transferred "as is" without adaptation. On this shallow portion of the demand curve for transferable technology, the cost is constant and relatively low. The downward slope of this flatter segment is due entirely to the diminishing returns of applying knowledge. Beyond the backlog, the demand curves slope downward rapidly, representing the added force of increasing costs to produce new knowledge.

At the rate of interest $r_1$, $I_1T_1$ is the total expenditure for knowledge acquisition, $I_1O$ for indigenous research, and $OT_1$ for acquisition from the existing backlog. Shortly, we will show that the true demand curve for indigenous research lies further to the left, although the expenditures will still be a marginal part of the total spending on knowledge acquisition. The initial conclusion corresponds to what is considered realistic when one accepts the notion of a cheap, sufficient, and easily digested backlog. A closer look at reality, however, forces one to consider

FIGURE 1

Indigenous and Foreign Knowledge Acquisition by LDCs

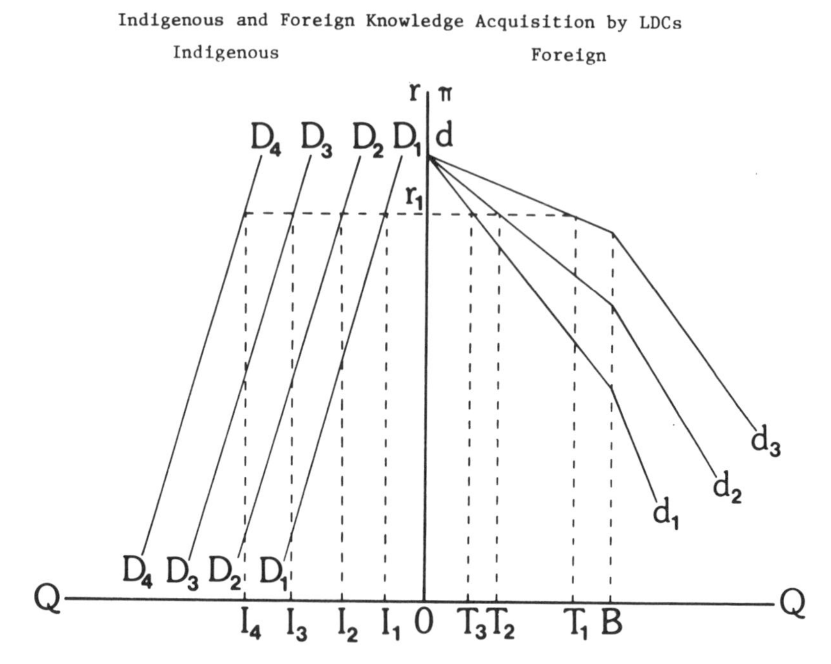


three factors: (1) the composition of the backlog relative to the needs of LDCs, (2) the costs of transferring and applying knowledge from MDCs and (3) the returns to indigenous research in LDCs. One must also pursue the policy implications of these considerations and the broad socioeconomic effect of indigenous research in LDCs.

*Backlog Insufficiencies*

Perhaps the most obvious shortcoming of the simplified model presented in Figure 1 lies in the fact that there are gaps in the existing backlog of knowledge obtainable from the MDCs. In many instances, deficiencies in the existing backlog are associated with specific geographical or environmental characteristics of developing regions. Illustratively, Evenson, Houck, and Ruttan investigated the transferral of new varieties of sugarcane and found that the breeding and development of the new varieties, in order to be compatible with specific soil, climate, and disease conditions and cultivating techniques of a particular geographic area, had to be undertaken by experiment stations located within the subregions.[16] A comprehensive survey of the Mexican science and technology situation pointed out the need for localized research on agricultural output in poorly rain-fed zones, the pharmaceutical potential of marine organisms, marine resources in general, the spread of particulates and gas in the atmosphere, and weather modification.[17]

Research on locally available materials also ranks high in importance. Construction techniques utilizing locally available material hold great promise, especially when they are associated with new building techniques that cut costs and increase employment.[18] Techniques that extend or complement traditional methods are particularly useful. Investigation is also needed to help lower production costs and expand uses for raw commodities. Mahbub Ul Haq complains that western countries are conducting a search for synthetic substitutes for jute, while specific knowledge on how to increase efficiency in jute production is not available from the same sources.[19] Other examples are developmental research on use of natural organic products as wood adhesives, replacing synthetic phenolics;[20] research reviving the use of Mexican plants in steroid hormone products;[21] and the development of alcohol fuel from sugarcane and other tropical plants.[22]

The present technological backlog in the advanced countries is inadequate to handle the reduction of waste or the productive use of waste products in LDCs. Bio-gas technology can simultaneously increase energy sources, reduce health hazards, and furnish more fertilizer.[23] In India, it is estimated that food valued at $3 billion is lost in the fields because of deficiencies in storage and transportation, a condition begging for applied research.[24] The same source also points out that the lower glass and metal content of village garbage in India renders it more conducive to bio-gas production than is most refuse in the MDCs.[25]

Economic conditions in LDCs frequently call for technologies not conveniently available from the industrial countries. Research to develop smaller-scale processing units more suitable for the market sizes of low-income countries needs to receive more attention. "Scaling down" of chemical equipment has met with some success in Mexico.[26] A small paper plant utilizing agricultural wastes producing ten tons per day has been recommended,[27] as has a reduced-scale cement plant.[28] Small skid-mounted ammonia/urea plants producing for limited local fertilizer markets may also be feasible.[29]

Special socioeconomic conditions demand different research directions as well. Complaints are heard in LDCs about the lack of medicines to combat parasitic or infectious communicable diseases and of pharmaceutical products suited to the lower incomes of the general population.[30] For example, in Brazil only about 30 percent of the population participates in the commercial pharmaceutical market.[31]

This thin sampling of cases is far from complete; literally dozens more are cited by the CONACYT, UNIDO, UNESCO, and UN.[32] How does this influence our analytical model? Originally the assumption was made that the backlog, while not infinite, did include an array suitable to LDCs. Obviously, in light of the evidence, we must now drop this assumption. The effect will be to increase the number of indigenous research projects that are technically and economically feasible. If we assume that these projects have expected internal returns ranging from very high to very low, the number of projects associated with each level of return will increase, symbolized by a shift in the demand curve for indigenous research outward to $DD_2$ and by expenditures rising to $I_2$ (Figure 1).

*Costs of Transfer*

The simple model has thus far rested on the assumption that knowledge is available at relatively low and constant costs in the MDCs. This would be fairly realistic for nonproprietary knowledge available in the form of books, journals, films, or instructional manuals. Even though they are inexpensive, such materials are subject to limitations. Business executives in Chile, for instance, believe that this type of information is sound for basic training of technicians and for professionals at universities, but books, reports, and handbooks are no substitutes for foreign technicians or study abroad when it comes to the more practical aspects of using novel technologies.[33]

Even though a portion of the cost of sending students abroad is defrayed by host government contributions to education in general and aid to foreign students specifically, the net expenditure by the LDCs is not negligible. Importing foreign technicians can also be extremely expensive. The going rate in Pakistan during the early 1970s was about forty dollars per day and has, no doubt, increased since.[34] Suri reports that, in India, wage differentials, costs of travel, and other allowances paid to foreign consultants raised their cost eight to ten times over that for local experts.[35] Even when the services of technical assistants are free to the recipient country, they usually absorb resources in the form of office space, housing, transportation, and local supplies. Perhaps most costly of all, "in many less developed countries senior officials and ministers spend a good deal of their time briefing foreign 'experts' on the problems of their country."[36] Thus, in reality these *direct* costs can be more formidable than our simplified model implies. There are *indirect* costs associated with the transfer of scientific and technical knowledge that seldom enter into the calculus of the acquiring agency.

If there is any generalization that can be derived from the literature on which there is near unanimity, it is the proposition that in virtually every case of transfer, *some* adaptations must be made. Differences in altitude, rates of corrosion, temperature, electric voltages, tariffs on spare parts, availability of repair facilities, availability and quality of raw materials, laws affecting shift work, degree of commitment of the labor force, range of labor skills, size of the market, prices for productive resources, and a host of other conditions will necessitate what some have called "creative adaptation." Yet, these costs are frequently ignored or underestimated. "In fact, the task of adaptation always remains, but is often underestimated because everything *appears* to be so cut and dried."[37] As we shall see, costs of adaptation are usually recognized by social scientists, but not fully taken into account by the individual or agency actually importing the knowledge.

These adaptations usually involve resources—particularly human resources—that are extremely scarce in less developed countries. One observer points out that adaptations require "creative, scientific imagination of a high order."[38] Similarly, Nathan Rosenberg speaks of some adaptations that require so much talent and originality that the term "innovation" should supplant "transfer" or "diffusion."[39] One international agency has called the need for adaptation a "wastage or devaluation of technology."[40] Certainly the social opportunity cost of resources thus employed is high and increases the real cost of transferring knowledge to less developed countries.

To assume that existing technology in MDCs can be effectively used "as is" in less developed countries is to ignore an additional cost. This is the cost of resources used to adapt the technology *plus* a residual cost involving the inability to adapt perfectly. The writer can think of no better example than the well-documented condition of widespread overcapacity in the industrial sectors of LDCs.[41] Undoubtedly, some of the causes are institutional measures that encourage the intensive use of capital, and others are attributable to the imperfections of the transfer mechanism and to lack of effort to scale down operations. To illustrate, a Nigerian study finds that in the purchase of turn-key factories from MDCs, the buyers give little or no consideration to the appropriateness of scale of the factories.[42] Jack Baranson has provided another Nigerian example describing a feasibility study of the plastics industry. The choice of technology was confined to large-scale equipment requiring imported raw materials. Baranson complains that equipment should have been considered that could produce a variety of plastics for small markets and that would employ processes using locally available inputs.[43]

Aside from buying technology embodied in productive hardware, businessmen in the LDCs are strongly tempted to accept disembodied technology associated with capital equipment in MDCs. Sales promotion, distribution, assembly line organization, job or task assignments, etc., often come along with the plant. Such culture-bound managerial techniques may be ill-suited to the conditions in less developed countries and lead to inefficiency. As a case in point, several U.S.-managed firms with plants in the Mexican Border Industrialization Program have insisted that foremen punch in on time cards. This stipulation ignores the educational and attitudinal characteristics of first-line supervisory employees in Mexico compared with similar workers in the United States. Many Mexican supervisors are college graduates and look upon themselves as part of management, with the capability of operating effectively without close managerial controls. As a result, there is a loss of morale, and most of the "clock punching" foremen put in an eight-hour shift whereas they had previously come early and voluntarily worked one or two hours overtime.

It seems reasonable to assume that the greater the proportion of indigenous technology that goes into developing the core productive hardware, the more likely the organizational techniques surrounding it will reflect existing cultural and institutional conditions in LDCs.

Another transfer cost takes the form of reduced efficiency as each "bundle" of knowledge is transferred. W. Paul Strassmann views the network for transferring information as a multistation relay network.[44] Information flows in both directions—some information is filtered out, and some information is added at each station. For lack of time and because of divergent interests, as information flows among machinery suppliers, consultants, trade journals, government advisors, etc., the quality of the information tends to deteriorate. Put in economic terms, the marginal productivity of the information declines roughly in proportion to the number of relay stations involved. Information is perishable due to excess handling. Conceivably, indigenous technology in LDCs could operate more effectively by reducing the number of relay stations and by developing a more homogeneous set of interests in effecting transfers. Perhaps this is why an Indian study stressed the need for continuity of research and development activity from the laboratory stage to the commercial exploitation of a new development.[45] The study held that close-knit communications may be more easily attained in a private research institute or in a private firm. The deterioration of information through the relay system can be thought of as a rise in the cost of transferring knowledge in the sense that the cost of importing *effective* knowledge is increased.

Another formidable increase in costs of knowledge to LDCs can be attributed to the fact that (a) much knowledge is proprietary and that (b) a seller's market exists for knowledge. Buyers bargain from a position of incomplete knowledge, limited alternative sources, and often must obtain new knowledge as part of a package that includes capital equipment, other production inputs, managerial talent, and international marketing expertise. Agreements for transfer of technology have often manifested these market advantages in the form of restrictive clauses calling for the prohibition of exports, tying purchase agreements, provisions stipulating that product or process improvements made in LDCs revert to the parent firm, and restrictions on the volume of output.[46] In addition, excessive prices charged to foreign affiliates in order to lower profits reported in LDCs are detrimental to the balance of payments, reduce the taxes collected on foreign-connected activity, and discourage the purchase of locally produced inputs in poor countries.[47] These advantages to the seller are intensified by differentiating buyers according to their willingness and ability to pay.

A closely related situation may also be considered as a gap in the backlog of knowledge. It entails instances in which firms in MDCs refuse to sell knowledge at *any* reasonable price. From 1950 to 1955 the Indian government contacted practically all of the manufacturers of optical glass in the United States, United Kingdom, France, Germany, and Japan to secure technical assistance. No agreement for collaboration could be reached, even though amounts exceeding one million rupees as a fee were offered, excluding the cost of equipment.[48] The choice was reduced to developing indigenous methods or relying permanently on imports of optical glass.

To summarize, aside from the *direct* costs of transferring knowledge, a closer scrutiny has uncovered *indirect* costs in the form of resources allocated to adaptation, inability to adapt fully, a tendency to adopt disembodied knowledge "as is" along with hardware, a loss of efficiency in the international information transfer network, and various monopoly effects. The analytical model can accommodate these indirect costs by showing two shifts. The internal rates of return on a range of bundles of knowledge available in MDCs are lowered because of higher effective costs, thus shifting the demand curve for foreign technology downward to $dd_1$ (Figure 1). Also, the bundles that *are* feasible to import are associated with a variety of adaptive research and development activities that must be undertaken indigenously; this is depicted as an increase in the demand for research and development projects in LDCs to $DD_3$. The amount of expenditure on indigenous research is now shown by $OI_2$ in the LDCs panel; that on transfer is shown by $OT_2$ in the foreign panel.

*Underestimated Gains*

There are a number of social returns that may accompany indigenous research and development that are usually not taken into consideration when assessing the feasibility of individual research projects. Reducing the "brain drain" of skilled scientists and technicians from LDCs offers one such possibility. Emigration of skilled people from LDCs is a result not only of higher salaries but also of better research opportunities. A survey conducted at the Indian Institute of Technology at Kanpur of individuals who worked abroad showed that greater opportunity for research ranked as the leading reason for their leaving India.[49] The added reward to indigenous research is calculated by computing the discounted present value of scientists and technicians who are induced to stay or to shorten their periods of absence and subsuming this value under domestic demand for research. It may well be, however, that financial inducements to remain at home become influential only after some critical minimum of local research facilities has been reached. It may also be pointed out that while LDCs will benefit, this will not necessarily increase *world* efficiency.

There is convincing evidence that indigenous research and technological borrowing are strong complements. Individuals who have had experience in coping with a particular set of local problems will be in a better position to participate in the transfer of knowledge. Indeed, they will be in a better position to ask the right questions.[50] And they will be better equipped to select appropriate techniques. "Only a research-minded country will be able to appraise the quality of technical know-how to be purchased, and the price to be paid for it. Indigenous research activity is thus very essential to economic development even if it does not immediately result in technical know-how."[51] Finally, there will be less tendency for an experienced researcher to become overawed by modern science and technology.[52] A classic example of what can be done is found in the Japanese "antennae" who were sent abroad to search for new technology.[53]

Indigenous research also has the complementary effect of reducing monopoly rents. Greater knowledge enhances the ability to "shop around" as well as the *potential* indigenous production of knowledge, based on a better bargaining position by LDCs. The complementary effects of better knowledge and lower monopoly costs increase the internal rate of return on both indigenous and transferred knowledge.

LDCs usually suffer large balance-of-payments deficits associated with the acquisition of technology. In the mid-1960s, LDCs paid 8 percent of the world total for the international transfer of technology and earned less than 1 percent.[54] Mason and Masson estimated that Latin America's net deficit on technology payments exceeded $300 million in the late 1960s.[55] By 1968, the cost of importing technology "already accounted for nearly 40 percent of the foreign official debt of developing countries, and 60 percent of direct investment by developed countries; and equaled the amount of multilateral aid."[56] The excessive transfer pricing practices of multinational firms mentioned above significantly increase the balance-of-payments drain.

There is nothing intrinsically undesirable about a payments imbalance for any single economic sector or activity. However, since many economists are convinced that the actual social opportunity cost of foreign exchange is above the price LDCs pay, and many believe that the restraint of scarce foreign exchange is a more formidable barrier to growth than a shortage of internal savings,[57] any progress in reducing the deficit that occurs as a by-product of achieving other justifiable goals must count as an added benefit.

Any reduction in the "brain drain" may somewhat reduce foreign exchange expenditures for training, although, presumably, most skilled individuals who go abroad eventually become self-supporting. The capacity to substitute domestic scientific and technical experts for imported experts will directly conserve foreign exchange. There may be an indirect effect as well, since evidence exists that foreign experts tend to be prejudiced against the use of local materials or techniques. One recent study has compared the relative amounts of domestic and imported steel required by three projects designed in Chile and three similar projects designed abroad. Chilean designs called for 22,000 tons of Chilean and 4,000 tons of imported steel; the foreign-designed projects called for 6,000 and 22,000 tons, respectively.[58] In India, local consultants reduced a proposed investment for tractor manufacturing from $13 million to $8 million and cut equipment import needs by $2.5 million.[59] In cases where indigenous technology encourages exports of local products or provides substitutes for imported proprietary know-how, a more obvious saving of foreign exchange will follow.

Another potential source of return to indigenous research results from the phenomenon of autonomous discoveries and inventions, which are sometimes a "fallout" of research directed to other ends. Accidental or incidental discoveries are quite common in the history of science and technology. Some anthropologists, sociologists, and economists have grasped the implications, but the majority of economists tend to regard such fallout as aberrations of little importance.[60] My own belief is that autonomous inventions and discoveries are much more important than orthodox economists are willing to admit. Such innovations create an extremely rewarding external economy, provided there is a large enough research apparatus to internalize or capture the new incidental discoveries. Much of the knowledge that is transferred to LDCs is in a "sterilized" form already embodied in material inputs, capital equipment, or pure theory. Most of the external economies of autonomous discovery will have been reaped in the original foreign research process.

Is indigenous research necessarily redundant in view of discoveries already made? The literature is replete with warnings about the inadvisability of spending precious funds to rediscover what others have already discovered. For the moment, let us put aside the consideration that *some* knowledge already known may be discovered cheaper than buying it and focus on a more fundamental misconception that such warnings support. If one speaks of pure science, then a repetition of research procedures will yield identical or very similar results. If we are speaking of technology, however, there is every likelihood that the end-product will be different. In the early development stage, or even in the latter stages of basic research, the path of the research thrust is markedly affected by the economic environment. For example, in MDCs, when two dozen designs are *technically* feasible, those not pursued will usually be those considered too expensive in labor costs for application. Relative factor prices do exercise a strong selective influence on knowledge production, and this influence reaches back to the earlier stages of research effort. Thus, in different economic environments, technological research and development are unlikely to lead to identical end-products; one would be surprised to find many instances of exact duplication.

A greater emphasis on indigenous research may reap *internal* economies as well. One dimension is found at the ultra-micro level within individual institutions. Many of the problems with indigenous research in less developed countries lie in the lack of inducements or pressures to overcome inertia within inadequately supported research institutes. Libraries are typically poorly organized. All too often scientists are subjected to petty regulations and controls. Little attention is given to the truly important routine of keeping detailed records on experimental results. Perhaps worst of all, promotions and administrative positions frequently depend on political skills rather than on research or administrative talents. What is needed is a properly motivated technocracy. The tasks that a technocracy does best, at least at the micro-level, are to take care of details that seem useful and to be intolerant of gross inefficiency. When the skilled technocracy becomes sufficiently large, influential, and vocal, scientific frauds cannot survive, and a purely bureaucratic approach to scientific effort will not be countenanced. This process depends partly, although not completely, on the magnitude of the local scientific and technological effort.

Additional economies of scale can be achieved at both the intrainstitutional and the macro-levels. Bath and James distinguish four critical thresholds that are useful conceptually, although difficult to apply empirically. (See Chapter 2.) As the absolute size of the science and technology effort increases, cross-fertilization occurs, linkages are formed, and interstices are filled.

Finally, imported technology is rarely associated with production of a wholly *new* product for the international market. Indigenous science and research are evidently needed if less developed countries are going to be able to engage in true product innovation. These potentials are enhanced by the possibility of exporting new capital goods to other less developed countries with similar conditions, or new consumer goods to countries with similar tastes and incomes. Yugoslavia, for instance, has been able to export technology to Egypt and Indonesia.[61] A United Nations investigation yielded the recommendation that Nigeria and other African nations make use of imported technology developed in Mexico and India.[62] Under present conditions, LDCs usually get in on the tail end of the product cycle as "late latecomers."

In summarizing this discussion of the returns that increased indigenous research encourages, there are a variety of rewards that may accrue to society that would not be taken into full consideration on each individual research venture. When such social benefits are taken into account, the internal rate of social return on the range of technically feasible indigenous research projects is increased. This has the effect of shifting the demand curve presented in Figure 1 to the left in the less developed countries' panel to $DD_4$; the optimal amount of indigenous research rises to $OI_4$. Two forms of return—(a) more efficient transfer of knowledge, and (b) a decline in monopoly rents—are complementary to the transfer process. The better selection and lower costs associated with transferred technology increase the internal rate of return on these bundles of knowledge also. Demand for the bundles of knowledge in the backlog thus increases to $dd_2$ in the foreign panel, and the expenditure on transferring knowledge moves to $OT_2$.

## Policy Recommendations

There are some scholars who believe that indigenous research and development resources are so inefficiently allocated, and performance is so poor, that a more sensible policy would be to improve the transfer mechanism. Ul Haq considers expenditures by Pakistan on developing nuclear energy as a gross misallocation.[63] There is also general agreement that government-sponsored research in India produces results far below what should be achieved.

Two comments are in order. First, there are many research projects in LDCs that have been outstandingly successful. The Technology Consultancy Centre of Kumasi in Ghana has found ways to use spent brewery waste for cattle feed and to manufacture soap from local inputs;[64] India's research laboratories dealing with metallic ores and low-grade coal have been successful,[65] as have her efforts to upgrade the leather industry. The Small Industry Extension Training Institute in Hyderabad has developed a multipurpose food additive from a mixture of peanut, sesame, and bengalgram.

The Mexican firm Productos Quimicos Vegetales is exploring industrial uses of national flora, the Instituto Mexicano para el Estudio de Plantas Medicinales is undertaking research on yucca and other plants,[66] and the Instituto Mexicano de Investigaciones Tecnológicas has developed a stabilized tortilla flour. Many of the other research activities mentioned earlier have also demonstrated their usefulness, such as those resulting in small-scale cement and paper plants, scaled-down chemical equipment, bio-gas production, and innovations in construction.

Second, the proper reaction to the Ul Haq position is: by all means improve the transfer mechanism! Although this analysis has attempted to show that more indigenous research is warranted, no one questions that the preponderant amount of knowledge will still be obtained from MDCs. Less developed countries may improve the conditions of transfer by international agreements outlawing tying agreements, governmental purchases of licenses and patents, patent pools, and development of special educational facilities in MDCs more appropriate to the needs of the recipient countries.

But what measures will encourage and improve indigenous research? Several concrete recommendations follow:

(1) It is becoming evident that an array of technologies appropriate to LDCs' conditions is available, and more techniques are constantly being developed. Furthermore, the feasibility and incidence of adaptation of appropriate technology are sensitive to factor prices and to the patterns of distribution of income that affect consumer choices.[67] A prerequisite to a concerted and large-scale indigenous effort to promote the development of such technologies is for LDCs to reduce imperfections in

their own factor markets and the degree to which income distribution favors the few.

(2) Some indigenous research projects in LDCs appear to have extremely attractive returns.[68] Also, some research activities clearly favor the comparative advantage of LDCs.[69] According to James, Jedlicka, and Street, priorities should go to (a) projects that hold promise of high economic return per unit of cost, either because of intrinsic merits or because of unusually severe monopolistic elements in the transfer mechanism, (b) projects that contribute to achieving or sustaining a critical mass of researchers in a particular endeavor, (c) projects that are sufficiently flexible to be able to be used in alternative ways in case of failure to attain the originally designated objectives, and (d) projects that will significantly absorb additional labor.[70]

(3) It is imperative that public research agencies be evaluated or, when appropriate, put to some empirical test. Inefficient institutions and projects should be eliminated, yet autonomous government agencies that have been extremely important in research and development activities in LDCs rarely go out of existence on a volunatry basis when they are no longer productive. Some mechanism for "creative destruction," to use Schumpeter's term, is necessary. To the extent that investigations are carried out for commercial ends, they should be subjected to a market test. If commercial earnings are not feasible, but significant social returns are claimed, a careful cost-benefit study of results should be conducted by an outside agency. Research projects of a more basic or fundamental character should be subject to preliminary and progress reviews by a competent outside board that includes distinguished scientists. It may be surprising how easily intra- and inter-agency equipment exchanges can be arranged, equipment maintenance carried out, and customs-free imports of research equipment worked out with the proper pressures and incentives.

(4) For the most part, successful institutions should be given first claim on limited research funds. Generally speaking, these are institutes associated with agricultural products or raw materials in LDCs. "Going with success" has the advantage of investing in established fields and reduces the risk of sustaining poor returns. This does not, of course, imply that some resources should never be devoted to exploring untried research areas.

(5) Evaluative investigations of research institutions should be made on intra- and inter-country bases. In particular, highly successful institutions should be examined as models. A variety of questions arise; e.g., how is the wedding of pure and applied research achieved in light of the well-known difficulty of doing so in LDCs? What are the

organizational requirements of the research? In what other countries and on what types of research could such an approach be successfully developed?[71]

(6) According to Alexander King, LDCs should specialize in their own internal research to take advantage of scale or "critical minima" effects,[72] but a greater degree of breadth can be achieved by selective participation in international scientific and technological projects. Intra-Third World cooperation also increases the bargaining power of LDCs and decreases monopoly profits on proprietary knowledge.

(7) Some institutional arrangements, very likely multilateral arrangements, must be made to ameliorate massive and almost routine political intervention in scientific and technological programs in Third World nations. Argentina provides an extreme example of frequent intervention. Militarists who gained power in 1966 purged teachers, including scientists, from the universities; Peronists purged conservatives after 1973; and after March 1976, there was a particularly pronounced repression of leftist and moderate professionals.[73] The disruption to scientific and technological experiments is severe, as is the damage done to the morale and commitment of the researchers.

(8) A systematic and concerted effort should be made to achieve an earlier entry into the product cycle. As conceived by Raymond Vernon, the typical product originates in a mature industrial country as the result of research and entrepreneurial activity. The product matures when it spreads to other industrial countries and finally becomes standardized when it is a candidate for production in virtually any country. With greater predictive power and some early adaptation of the original production process, it may be possible to enter an earlier stage of the product cycle rather than wait for the standardized information to become available, with a high price tag attached.

(9) LDCs should be encouraged not to neglect seemingly minor and pedestrian areas for research. Research on methods of maintenance and repair, local fabrication of spare parts, and the design of multipurpose machine tools, while failing to conjure up images of a shiny, modern scientific laboratory, are nonetheless extremely important. The design of a small inexpensive refrigeration unit may reduce the enormous quantity of premarket food spoilage in LDCs. Similarly, "soft sciences" should reinforce research conducted along more traditional lines. Such studies may improve teaching techniques, encourage pooling small amounts of personal savings, and enhance the managerial capabilities of small enterprises.

While the foregoing constitute concrete recommendations, I would like to add a general observation. A new perspective on the process of economic development is needed—one which places technological advance in a central focus. My collaborators and I believe that American institutionalism, as developed by Thorstein Veblen and C. E. Ayres, forms a valuable counterbalance to the dependency explanation of underdevelopment that holds sway in Latin America and is gaining in popularity in other Third World regions.[74] Institutional economists share with dependency theorists a dissatisfaction with the relevance of standard economics to developmental problems of poor countries; both groups believe that economists cannot avoid making value judgements, and both use a broad multidisciplinary approach. Institutionalism, however, is less encumbered with Marxist ideology and criticizes obstructive institutions whether they be transplanted or domestic. It emphasizes continual improvements over a broad front within an evolving continuum of goals and values; there is no need to wait for disengagement from the wealthier countries (as some extreme *dependentistas* advocate) before positive action can be taken. Moreover, institutionalism centers on technological advance as a potentially autonomous force, as the motor of the process of development. Third World social scientists and administrators interested in science and technology policy could benefit from the insights that an institutionalist perspective provides in considering the long course of technological progress.

**Notes**

1. Moses Abramovitz, "Resource and Output Trends in the U.S. Since 1870," *American Economic Review* 40, no. 2 (1956):1–23; John W. Kendrick, "Productivity Trends: Capital and Labor," *Review of Economics and Statistics* 38, no. 3 (1956):248–57; Robert M. Solow, "Technical Change and the Aggregate Production Function," *Review of Economics and Statistics* 39, no. 3 (1957):312–20; Edward Denison, *The Sources of Economic Growth in the U.S.* (New York: Committee for Economic Development, 1962).

2. United Nations, *Conference on the Application of Science and Technology for the Benefit of Less Developed Areas,* 12 vol. (New York: UN, 1963); Richard L. Meier, *Science and Economic Development: New Patterns of Living* (Cambridge, Massachusetts: MIT Press, 1966).

3. Dilmus James, "Bibliography on Science and Technology Policy in Latin America," *Latin America Research Review* 12, no. 3 (1977):71–101.

4. Jorge A. Sábato, "The Influence of Indigenous Research and Development Efforts on the Industrialization of Developing Countries," *Industrialization and Development,* ed. H. E. Hoelscher and M. C. Hawks (San Francisco: San Francisco Press, 1969), pp. 178–83; Diana Crane, "An Inter-organizational Approach to the Development of Indigenous Technological Capabilities: Some Reflections on the Literature," mimeographed, Occasional Paper no. 6, CD/TI (75) 10, Organization for Economic Development and Cooperation, 1974; M. M. Suri, "Local Capability and Preparedness for Appropriate Technology Transfer to Developing Countries," World Employment Program Working Paper WEP 2-22/WP. 27, International Labor Office, 1975; Dilmus James, Allen Jedlicka, and James Street, "Issues in Indigenous Research and Development in Third World Countries" (Paper delivered at the American Association for the Advancement of Science, Denver, Colorado, February 1977); idem, "The Relevance of Institutional Economics to Science and Technology Strategies of Latin America" (Paper delivered at Latin American Studies Association, Houston, Texas, November 1977).

5. Marcel Roche, *BITACORA—1963,* Ediciones IVIC (Caracas, Venezuela, 1963), p. 188.

6. Olga Gasparini, *La Investigación en Venezuela; Condiciones de su desarrollo,* Publicaciones IVIC (Caracas, Venezuela, 1969), p. 84.

7. Society for International Development, "Rio Report Proposes Structural Changes in Global System to Help Obtain by Year 2000 a More Equitable International Social and Economic Order," *Survey of International Development* 8, no. 6 (1976):3.

8. Hans Singer cited in Edgar O. Edwards, ed., *Employment in Developing Countries* (New York: Columbia University Press, 1974), p. 26.

9. M. M. Suri, "Local Capability for Appropriate Technology Transfer," p. 3.

10. D. Babatunde Thomas, *Capital Accumulation and Technology Transfer: A Comparative Analysis of Nigeria Manufacturing Industries* (New York: Praeger Publishers, 1975), p. 43; Maximo Halty Carrére, *Situación actual para el desarrollo científico y técnico. Implicaciones al nivel de política de estrategía* (Washington, D.C.: Organization of American States, 1969).

11. Ignacy Sachs, "Selection of Techniques: Problems and Policies for Latin America," *Economic Bulletin for Latin America* 15, no. 1 (1970):22.

12. Economic and Scientific Research Foundation, *Research and Industry: Seven Case Histories* (New Delhi: ESRF, 1966), pp. 1, 20.

13. Francisco Sagasti and Alberto Araoz, *Science and Technology Policy Implementation in Less Developed Countries: Methodological Guidelines for the STIP Project* (Ottowa, Ontario, Canada: International Development Research Centre, 1976).

14. Kenneth Arrow, "Economic Welfare and the Allocation of Resources for Invention," in National Bureau of Economic Research, *The Rate and Direction of Inventive Activity: Economic and Social Factors,* (Princeton, N.J.: Princeton University Press, 1962), pp. 609–26.

15. The same spectrum of uncertainty can, of course, be applied to the supply side. For a case of cost estimates exploding at a tragicomedy rate (but with a happy ending), see the development of the first radio telescope (Bernard Lovell, *The Story of Jodrell Bank* [New York: Harper and Row, 1968]).

16. R. E. Evenson, J. P. Houck, and V. W. Ruttan, "Technical Change and Agricultural Trade: Three Examples—Sugarcane, Bananas, and Rice," *The Technology Factor in International Trade,* ed. Raymond Vernon (New York: Columbia University Press, 1970), p. 421.

17. Consejo Nacional de Ciéncia y Tecnología, *National Indicative Plan for Science and Technology* (Mexico, 1976), pp. 121, 126, 189, and 195.

18. See the Strassmann contribution to the present volume for several examples.

19. Mahbub Ul Haq, "Wasted Investment in Scientific Research," *Science and the Human Condition in India and Pakistan,* ed. Ward Moorhouse (New York: Rockefeller University Press, 1968), p. 128.

20. E. Kulvik, "Review of Past Research on Utilization of Naturally Occurring Organic Products as Replacement of Synthetic Phenolics in Wood Adhesives," ID/WG. 248.2 (United Nations Industrial Development Organization, 1977).

21. Gary Gereffi, "Drug Firms and Dependency in Mexico: The Case of Steroid Hormone Industry," *International Organizations* 32, no. 1 (Winter 1978):237–86.

22. William Hieronymus, "Brazil Tries Mixing Alcohol from Sugar with Gasoline to Reduce Its Oil Imports," *Wall Street Journal,* Southwestern Edition (November 28, 1977), p. 22.

23. Allen Jedlicka, "Una fuente barate de combustible para la población rural latinoamericana," *Interciencia* 1, no. 1 (1976):45–46.

24. M. M. Suri, "Local Capability and Preparedness for Appropriate Technology Transfer to Developing Countries," p. 13.

25. Ibid., p. 18.

26. José Giral Barnes, "Development of Appropriate Chemical Technology in Mexico," *Choice and Adaptation of Technology in Developing Countries: An Overview of Major Policy Issues* (Paris: Organization for Economic Cooperation and Development, 1974), pp. 182–86.

27. M. M. Suri, "Local Capability for Appropriate Technology Transfer," pp. 10–12.

28. H. D. Visvesvaraya, "Improved Packaging for Cement and Mini Cement" (ID/WG. 246/1), mimeographed, (United Nations Industrial Development Organization, 1977).

29. World Bank, *Appropriate Technology in World Bank Activities* (Washington, D.C.: World Bank, 1976), p. 85.

30. United Nations Industrial Development Organization, "The Establishment of Pharmaceutical Industries in Developing Countries," mimeographed (Vienna: UNIDO, 1969), p. 13.

31. Carlos Osmar Bertero, "Drugs and Dependency in Brazil—An Empirical Study of Dependency Theory: The Case of the Pharmaceutical Industry," Dissertation series no. 36 (Ithaca, New York: Cornell University, Latin American Studies Program, 1975), p. 206.

32. Consejo Nacional de Ciéncia y Tecnología, *National Indicative Plan for Science and Technology;* United Nations Industrial Development Organization, "Cooperative Programme of Action on Appropriate Industrial Technology (ID/B/188), mimeographed (Vienna: UNIDO, 1977); United Nations Educational, Scientific, and Cultural Organization, *Science and Technology in Asian Development* (Paris: UNESCO, 1970); and United Nations, *World Plan of Action for the Application of Science and Technology to Development* (New York: UN, 1971).

33. Fernando Aguirre, Fernán Ibañez, and United Nations, Department of Economic and Social Affairs, Division of Public Finance and Financial Institutions, *Arrangements for the Transfer of Operative Technology to Developing Countries: Case Study of Chile* (ESA/FF/AC.2/13), mimeographed (New York: UN, 1971).

34. Investment Advisory Center of Pakistan and United Nations, Department of Economic and Social Affairs, Division of Public Finance and Financial Institutions, *Arrangements for the Transfer of Operative Technology to Developing Countries: Case Study of Pakistan* (ESA/FF/AC.2/18), mimeographed (New York: UN, 1971).

35. M. M. Suri, "Local Capability for Appropriate Technology Transfer," p. 15.

36. Angus Maddison, *Foreign Skills and Technical Assistance in Economic Development* (Paris: Organization for Economic Coopera-

tion and Development, 1965), p. 13.

37. Albert O. Hirschman, *The Strategy of Economic Development* (New Haven, Conn.: Yale University Press, 1958), p. 38.

38. C. R. Powell, "Priorities in Science and Technology for Developing Countries," *Science and Society,* ed. Maurice Goldsmith and Alan Mackay (New York: Simon and Shuster, 1964), p. 82.

39. Nathan Rosenberg, "Comment (to Evenson et al.)," *The Technology Factor in International Trade,* ed. Raymond Vernon (New York: Columbia University Press, 1970).

40. United Nations Educational, Scientific, and Cultural Organization, *International Aspects of Technological Innovation* (Paris: UNESCO, 1971), p. 84.

41. As an example, it is estimated that 30 to 40 percent of Mexico's industrial capacity is not used. (Miguel S. Wionczek and División de Hacienda Pública e Instituciones Financieras, "La transferencia internacional de tecnología al nivel de empresa: El caso de México" (ESA/FF/AC.2/10), mimeographed (New York: UN, 1971), p. 86.

42. United Nations, Division of Public Finance and Financial Institutions, *Arrangements for the Transfer of Operative Technology to Developing Countries: A Case Study of Nigeria* (ESA/FF/AC.2/4), mimeographed (New York: UN, 1971), p. 31.

43. Jack Baranson, "Economic and Social Considerations in Adapting Technologies for Developing Countries," *Technology and Culture* 6, no. 1 (1963):27–28.

44. W. Paul Strassmann, *Technological Change and Economic Development: The Manufacturing Experience of Mexico and Puerto Rico* (Ithaca, New York: Cornell University Press, 1968), pp. 29–31.

45. Economic and Scientific Research Foundation, *Reseaıch and Industry: Seven Case Histories.*

46. For a summary of these practices, see Dimitri Germidis and Christine Brochet, "The Price of Technology Transfer in Developing Countries," Industry and Technology Special Study no. 91 (CD/TI [75] 7), mimeographed (Paris: Organization for Economic Cooperation and Development, 1975).

47. Constantine V. Vaitsos, *Intercountry Income Distribution and Transnational Enterprises* (London: Clarendon Press, 1974).

48. Economic and Scientific Research Foundation, *Research and Industry: Seven Case Histories,* p. 52.

49. Marshal F. Merriam, "Brain Drain Study I.I.T. Kanpur," mimeographed (Kanpur, India: Indian Institute of Technology Kanpur, 1969), pp. 28–29.

50. Constantine V. Vaitsos, "Transfer of Resources and Preservation

of Monopoly Rents," Economic Development Report no. 168, mimeographed (Cambridge, Mass.: Development Advisory Service, 1974), pp. 16–17.

51. Economic and Scientific Research Foundation, *Research and Industry: Seven Case Histories,* p. 16.

52. Hirschman claims indigenous entrepreneurs suffer this fate from the demonstration effect of foreign enterprises (Albert O. Hirschman, "Cómo y por qué desinvertir en América Latina," *El Trimestre Económico* 37, no. 3, (1970):489–514).

53. Daniel Lloyd Spencer, *Technology Gap in Perspective* (New York: Spartan Books, 1970), pp. 88–89.

54. C.D.G. Oldham, C. Freeman, and E. Turkan, "Transfer of Technology to Developing Countries" (TD/28/Supp. I), mimeographed (Geneva: United Nations Council on Trade and Development, 1967).

55. R. Hal Mason and Francis G. Masson, "Balance of Payments Costs and Conditions of Technology Transfer to Latin America," *Journal of International Business* 5, no. 1 (Spring 1974):73–85.

56. Dimitri Germidis and Christine Brochet, "Price of Technology Transfer," p. 11.

57. Hollis B. Chenery and A. M. Strout, "Foreign Assistance and Economic Development," *American Economic Review* 56, no. 4 (1966):679–733.

58. Fernando Aguirre, Fernán Ibañez, and United Nations, Department of Economic and Social Affairs, Division of Public Finance and Financial Institutions, *Arrangements for the Transfer of Operative Technology to Developing Countries: Case Study of Chile,* p. 93.

59. M. M. Suri, "Local Capability for Appropriate Technology Transfer," p. 14.

60. See writings by both Ogburn and Ayres (W. F. Ogburn, *Social Change* [New York: B. W. Hurebsch, 1922]; C. E. Ayres, *Theory of Economic Progress* [Chapel Hill, North Carolina: University of North Carolina Press, 1944]).

61. Jack Baranson, "Economic and Social Consideration," p. 85.

62. United Nations, Division of Public Finance, *Arrangements for the Transfer of Operative Technology to Developing Countries: A Case Study of Nigeria,* p. 85.

63. Mahbub Ul Haq, "Wasted Investment in Scientific Research," p. 128.

64. John W. Powell, *Annual Review No. 4, 1975–1976,* Technology Consultancy Centre (Kumasi, Ghana: University of Science and Technology, 1976); idem, "An Intermediate Technology Rule for a

University in the Third World" (Paper delivered at the American Association for the Advancement of Science, Denver, Colorado, February 1977).

65. Economic and Scientific Research Foundation, *Research and Industry: Seven Case Histories,* p. 26.

66. Consejo Nacional de Ciéncia y Tecnología, *National Indicative Plan for Science and Technology,* p. 172.

67. For a case claiming that selection and adaptation of technologies would be sensitive to more rational factor prices, see the Bath and James contribution to this volume (Chapter 2). For corroborative evidence that elasticities of factor substitutions are above or close to unity when technological changes are included in the analysis, see A. Berry, "The Rate of Interest and the Demand for Labor," Center Discussion Paper no. 144, mimeographed (New Haven, Conn.: Yale University, Economic Growth Center, 1976); Jaime de Melo, "Effects of Distortions in the Factor Market: Some General Equilibrium Estimates," Discussion Paper no. 34 (Washington, D.C.: Agency for International Development, 1976) and René Villareal Arrambide, *El disequilibrio externo en la industrialización de México 1929-1973* (Mexico, D.F.: Fondo de Cultura Económica, 1976).

Gerard Boon, as cited by Strassmann, estimates the elasticity of substitution between labor and capital to be .973 in the construction sector, while Strassmann himself, referring to construction in Mexico, estimates the elasticity of substitution between building materials and labor to be around 1.2 (W. Paul Strassmann, "The Substitution of Materials or Capital for Labour in Mexican Construction," *Studies on Employment in the Mexican Housing Industry,* ed. Christian G. Araud [Paris: Organization for Economic Cooperation and Development, 1973], p. 308). Wayne Thirsk places the labor-capital substitution elasticity in Colombian agriculture at about 1.5 ("Factor Substitution in Colombian Agriculture," *American Journal of Agricultural Economics* 56, no. 1 [1974]:73–84). For the effect of income distribution on technology through influencing product choice, see Francis Stewart, "Choice of Technique in Developing Countries," *Journal of Development Studies* 9, no. 1 (1972):99–121, and case studies by C. Baron; C. Cooper et al.; and Stewart—all found in A. S. Bhalla, ed. *Technology and Employment in Industry* (Geneva: International Labor Office, 1975).

68. See Arndt and Ruttan for evidence relating to agricultural research. (Thomas Arndt and Vernon W. Ruttan, "Resource Allocation and Productivity in National and International Agricultural Research" [New York: Agricultural Development Council, 1975]).

69. Yash Pal, "Science and Development in India—Some Reflections," *Views on Science, Technology, and Development,* ed. Eugene Rabinowitch and Victor Rabinowitch (Oxford: Pergamon Press, 1975), p. 65.

70. Dilmus James, Allen Jedlicka, and James Street, "Issues in Indigenous Research and Development," p. 11.

71. Exemplary of such studies are the examination by Diogenes Hill et al. of the development of Colombian technology to produce texturized vegetable protein which was spearheaded by the Institución de Investigación Tecnológica ("Análisis preliminar de un fenómeno de innovación tecnológica de interés para el país: Introducción de las proteínas vegetables texturizadas en el mercado nacional" [Bogotá, Colombia: COLCIENCIAS, 1976]).

72. Alexander King, "Science International," *Science and Society,* ed. Maurice Goldsmith and Alan Mackay (New York: Simon and Schuster, 1964), p. 117.

73. These ideas are more fully developed in James H. Street, "Political Intervention in Scientific and Technological Activity in Latin America: An Unorthodox Proposal for Reducing the Harmful Effects" (Paper delivered at the American Association for the Advancement of Science, Houston, January 1979); Nicolas Wade, "Repression in Argentina: Scientists Caught Up in Tide of Terror," *Science* 194, no. 4272 (December 24, 1976):1397-99; idem, "Physics in Argentina," *Science* 196, no. 4296 (June 17, 1977):1302.

74. Dilmus James, Allen Jedlicka, and James Street, "Relevance of Institutional Economics to Science and Technology Strategies."

# Part 2
# Case Studies in Internal Technological Diffusion: Sectors and Countries

# 6
# Acquiring and Using Technological Information: Barriers Perceived by Colombian Industrialists

*Allen D. Jedlicka*
*Albert H. Rubenstein*

A compelling necessity in all developing countries has been the need to acquire relevant technology to improve the productive capacity of indigenous industry. Several alternative strategies have been suggested to meet this goal. Some believe that the problem is best left to the multinational companies, while others argue that the optimum solution is to improve or develop existing national facilities for technology transfer as well as the indigenous ability to absorb and apply available technology from other countries.[1] A statement by M. Halty Carrére defines the latter approach:

> An appropriate strategy for technical progress in any given country requires the proper combination of two types of policies:
>
> 1. The development at the national level of a well-balanced institutional system of research, education, training, and extension services which will permit the advancement of its own research capacity, the diffusion of its results, and the support of its application to the production system.
> 2. The selection, adaptation, assimilation, and utilization of the scientific and technological advances of other countries.[2]

Although this chapter is concerned with the effective utilization of technology, the primary topic of discussion has to do with the first policy enunciated by Carrére in regard to a study done in Colombia. The two policies are not independent of each other. Indeed, the development of a technological climate or environment that is receptive to technology and that creates its own demand for technology is an essential aspect of successful transfer within a country. In many countries, this climate does not exist. Neither does the means to create it through indigenous educational sources. It is toward this objective that an experiment in Bogotá, Colombia was conducted a the University of the Andes. The purpose was to develop a better climate for appropriate technology through the explicit training of engineering and business administration students in the mechanics and philosophy of such technology.

In addition to an improved climate, greater knowledge is needed on the perceived barriers to technology transfer. This was the central subject of the study described in this chapter.[3]

**Methodology of the Survey**

The primary task of the study was to develop a questionnaire which would (1) measure the predominant barriers to the acquisition of technological information and adoption of technology by Colombian industrialists, and (2) obtain some qualitative understanding of the ways in which industrialists obtained the technology needed to innovate within their companies.

A total of thirty-two companies, classified as large, medium, and small, was included in the sample.[4] No effort was made to isolate the sample to a specific industrial sector, and any company willing to respond to the questionnaire was included. Telephone inquiries to small companies typically indicated that they were essentially cottage industries that had never been involved in a transfer transaction with any agency. However, two small companies met the basic criterion (involvement with a technical diffusion agency) needed to respond to the questionnaire.

Managers who had recently made technological innovations were asked to respond to a list of thirty-one hypothetical barriers believed to affect the technology transfer process. These managers were asked to think consistently about each such innovation and to rate the barriers according to their relevance to the implementation of the innovation—as affecting the transfer very much, somewhat, or very little.[5] Informants were encouraged to describe their reaction to each barrier and its effect on the particular innovation considered. Additionally, their opinions were sought on the existing institutions for technology transfer in Colombia and on the possible results of introducing regional research and development centers in Colombia and neighboring Latin American Free Trade Association (LAFTA) countries. Analysis was limited to a simple numerical count of the most often cited barriers to transfer.

**Results of the Investigation**

As expected, some barriers were perceived to affect the technology transfer process significantly. These are listed below and are grouped into three categories: technological source barriers, internal barriers, and external barriers.

1. *Technological source barriers affecting adoption of technology*
    - Technological sources cannot provide us general information about new innovations that would be of interest to us.
    - Acquiring technological information from sources in this country (Colombia) is too slow.
    - Technological source extension agents lack experience and knowledge in the fields in which we need information.
    - The technological source extension agents lack the skill to judge our technical needs adequately.
2. *Internal barriers affecting adoption*
    - Implementation of innovations proposed by indigenous sources took too much time.
    - We couldn't tell (from the source's explanation) how the innovation would increase our profits.
    - The proposed innovation required too much maintenance.
    - The financial risk was too great.
3. *External barriers affecting adoption*
    - The potential market for products made by the innovation was too small and would not improve our market position.
    - Interest charges are too high to purchase the equipment needed for the innovation.

From the list above, one can see that ten of the thirty-one hypothetical barriers were perceived to be particularly significant by managers who had made technological innovations in all sizes of industrial companies.

In the first group of technological source barriers, industrialists found fault with the inadequate training of extension agents in the fields in which they needed information. This is a complaint which these men share with subsistence farmer technological recipients studied in the past and seems to be one of the predominant barriers in any effort to transfer technology in developing countries.[6] Industrialists felt that extension agents were not adequately trained to judge the technical needs of innovators; a view often expressed was that many persons playing the role of extension agent were trained in nontechnical areas totally unrelated to the industry's interests. This opinion is supported by another study, in which it was not uncommon to find lawyers attempting to fill the duties of extension agents.[8]

Slowness in acquiring information within the country was perceived to be due to bureaucratic difficulties in locating and obtaining information. Many of the managers, surprisingly, had developed their own efficient communication channels for information outside the country, most often with the United States. Obtaining information in this way was considered more reliable, accurate, and fast. One innovator in a family-owned company had sent his son to a U.S. university to obtain the engineering skills needed in the company. He insisted he had no need for technological source extension agents, since they were worthless to begin with and served largely as government agents to impose more taxes on his company. His son was also making a number of technological contacts in the United States to facilitate acquiring future information and technology after his return to Colombia.

The most significant barriers in the other two groups concerned the effects of implementation and maintenance costs of the prospective innovation, and financial and market risks in adopting an innovation.

Listed below are those barriers that were considered less significant than the preceding, but were found to affect the transfer process somewhat. Because of the very weak statistical support for these barriers, we do not suggest that they significantly affect the transfer process, but they do indicate some possible relationships, however tenuous.

1. *Technological source barriers affecting adoption of technology*
    - None of the technological sources considered how the technical change affects our future economic strength.
    - The technological sources are geographically too far away to be effective.
    - Very little technical information has been translated.
2. *Internal barriers affecting adoption*
    - We do not have the engineering staff to make the necessary technical changes.
    - The equipment required in the innovation was too complicated for us to handle.
3. *External barriers affecting adoption*
    - We could not obtain enough capital to purchase the innovation.
    - We would encounter union problems because the innovation would displace too many employees.
    - There are political problems involved in adopting the innovation (i.e., greater control and involvement of the government).

## Interpretation of the Results

As indicated above, several barriers were perceived to have a negative effect on the technology transfer process. Perhaps even more important were the responses to open-ended questions on the ways in which the transfer of technology and information services could be improved. When asked if a regional research and development (R&D) center, which would serve not only Colombian but other LAFTA countries, might be a viable solution, respondents generally replied that such a center could not function well because the member LAFTA countries would not fund it adequately. Most informants felt that even if the center were established, it could not meet all their technological needs and would become an ineffectual institution, useful only to the people who operated it by perpetuating their irrelevant livelihood.

Informants consistently expressed the feeling that existing technological sources in the country had nothing worthwhile to offer Colombian industrialists. As mentioned earlier, most managers spontaneously reported a perceived worthlessness of indigenous technological sources. In fact, an equally common feeling was that collaboration by the informants with indigenous technological sources who were predominantly publicly supported would be a liability because it would enable the government, through public servants in the technological agencies, to gain more information about the companies' operations and total income. A prevalent fear was that the result would be even greater control and higher taxes by government agencies. One extremely candid manager stated that he was already paying enough in bribes to officials, and the last thing he wanted to do was to extend the practice to new agencies.

Probably little can be done to overcome these major perceptual barriers to government technological sources until government agencies improve their existing facilities and modes of conduct with local industrialists (assuming the views of the latter informants are correct). Alternatively, government agencies must clear up misconceptions about the effectiveness of existing technological services and take measures to promote the transfer of technology to serve the interests of industrialists as well as the economic development of the country. As one step in that direction, the authors, in collaboration with several Colombian colleagues in the Department of Industrial Engineering at the University of the Andes in Bogotá, have tried to influence the training of engineering and administration students and advise government officials on the need for and the means of improving indigenous technological source institutions. This project is described below.

The first problem in discussing the production of technologies "appropriate" to a developing country is to agree on a definition of the term. The definition accepted and used by the project group is that an appropriate technology is one that effectively utilizes the manpower, resources, and institutional realities of a particular country.[8] These realities include such nontechnical aspects as the effects of government and labor union policy as well as indigenous entrepreneurial ability.

In the initial research design of the experiment, members of the project group felt that the key independent variable in an attempt to effect a significant change in the technological choice and development strategies of a developing country would be the training of engineering and administration students in a reputable indigenous university.[9] The resources of the university would also be used to influence government policy-makers and industrial entrepreneurs. In other words, concentrating on a long-term basis on the training of students and utilizing the information linkages of the university with the outside world would result in the effective development of appropriate technology in the country.[10]

The belief that the training of engineers and administration students will effect the desired change in technological development in Colombia rests on their internalized belief in the worthiness of appropriate technology. A necessary component of the training program includes having these potential policy-makers and researchers conduct basic research on variables that may impede or constrain the development of appropriate technology. These constraining variables were hypothesized to fit within three general problem-areas affecting the development of appropriate technology. Under the guidance of the Colombia project members, students investigated the following problem-areas in a research seminar, where it was expected that controls of these variables would be established by the students in their professional work:

1. Factors within the institutional technological framework that impede the development of appropriate technology.
2. Criteria utilized by entrepreneurs in industry for technology selection (whether appropriate or inappropriate).
3. Methodologies for implementing appropriate technologies in Colombian industry.

Assigning students to research has the benefit of teaching them the need for appropriate technology in their country; at the same time, the project group acquires information which will enable them to refine the experimental design, control constraining variables, and initiate other changes which can affect the long-term desired result of producing appropriate technologies in Colombia.

Because of the institutional connections of three of the members of the group, it was possible to utilize the University of the Andes as the institutional setting to modify the training of engineering and administration students. A grant from the Rockefeller Foundation was obtained to finance the first year's effort.

With the cooperation of the rector of the university and department heads, a course on "Technology Transfer and Development" was added to the engineering curriculum in the fall semester of 1973. It is now regularly offered each semester under the direction of one of the Colombian project members. The purpose of the course is to make engineering and administration students aware of the problems of technology transfer and to introduce them to the concept of appropriate technology as it relates to the development of their country.

The research seminar "Special Topics on Technological Policy, Choice, and Employment" serves as the principal vehicle for research on constraining variables affecting appropriate technology and, in turn, provides material for the technology transfer and development course. Over a period of years, it is expected that a reiterative feedback system of research and course modification will shed more light on the constraints. It is also hoped that the newly trained corps of engineering and administration students will understand and effect further progress in the organizations where they work on the development of appropriate technology.

## Recommendations Based on the Study

A positive effort to improve technological information services must be made to promote the technological independence of Latin American countries; yet exact replication of the modes for technology transfer used in the developed countries or duplication of all the relevant information from developed countries is not necessarily the answer.

For example, one way to overcome the limitations of an inadequate technological information system might be to link up Colombian or any other LAFTA institutes with counterpart institutes in the United States and Europe through satellite transmission. In that way, if the local institute did not have the information needed by a client, the counterpart institute could be called, and a printout of available technical abstracts could be obtained. If the printout met the client's interests, the full text of the information could then be sent airmail. On the other hand, the problem of improving the quality of extension services from domestic institutions to industry can be solved on the local level; educational training facilities exist in many countries and require only a commitment to fund such efforts by their governments. Some mechanism to maintain confidentiality might also help eliminate fears by industrialists that information acquired for technical purposes will be used for taxation.

To conclude with an optimistic comment on the process of technology transfer in Colombia, we believe that if a conscientious effort is made to improve the existing methods of information transfer within the country, Colombian industrialists will respond to the advantages of effective, apolitical, indigenous technological information and transfer services. The project at the University of the Andes may advance this process by educating and influencing the new breed of industrialists, technologists, and administrators who will make the necessary changes.

## Notes

1. Allen D. Jedlicka, "U.S. Multinational Companies and their Operations in LDCs," Summary in *India Forum,* May 1973.

2. M. Halty Carrére, "The Process of International Transfer of Technology: Some Comments Regarding Latin America," Department of Scientific Affairs, Pan American Union, Washington, D.C., 1968.

3. The study was a pilot effort of the OAS to develop regional dissemination centers for the LAFTA countries. Other participants in the design of the instrument were COLCIENCIAS in Bogotá, Colombia, and the Instituto Torcuato di Tella in Argentina. The data were collected by Maria Theresa Camargo, Enrique Ogliastri, and Allen Jedlicka.

4. We defined small companies as those with 25 or fewer employees, including management and workers; medium-sized companies had 25 to 250 employees; large-sized companies had more than 250 employees. In the sample, two companies were in the small category, twenty in the medium category, and ten in the large category.

5. Albert H. Rubenstein et al., "Standard Instruments and Procedures for Conducting National Studies of the Interaction between Sources and Users of Technical Information," Final report to the Department of Science and Technology, Organization of American States, Washington, D.C., October 1972.

6. Allen D. Jedlicka, "Comments on the Progress of an International Joint Study on Technological Choice and Employment in Developing Countries," Office of International Programs, National Science Foundation, Washington, D.C., March 1974.

7. A. Hoyos, "La Inversión Extranjera en Colombia: Un Estudio del Desarollo de Las Disposiciones Legales y Reglamentarias sobre la Inversión Extranjera anteriores al Año 1967" [Foreign Investment in

Colombia: A Study of Legal Regulations Prior to 1967], Working Document No. 8, Department of Industrial Engineering, University of the Andes, June 1974.

8. Members of the project group include Enrique Ogliastri, Carlos Dávila, and Jaime Silva of the Industrial Engineering Department at the University of the Andes in Bogotá, Colombia; Albert H. Rubenstein of the Industrial Engineering Department at Northwestern University, and Allen Jedlicka of the School of Business at the University of Northern Iowa.

9. "Experiment" is used in a rather broad sense. The dependent variable, the effective development of appropriate technology, is the result desired from the experiment. The primary independent variable is a change in the training of engineers and administrators who will play a more effective role in their country's development. Various intervening or parametric variables that affect the change cannot be controlled in the beginning. On a long-term basis, through continued research by students who ultimately will become technologists and policy-makers themselves, it is believed that these parametric variables can be both discovered and controlled.

10. Allen D. Jedlicka et al., "Factors Affecting the Transfer of Technology from Federal Agencies to State and Local Agencies," Proceedings of Midwest conference, American Institute for Decision Sciences, Indianapolis, April 1975.

# 7
# The Acquisition and Use of Technical Knowledge by Mexican Farmers of Limited Resources

*Allen D. Jedlicka*

The managerial aspects of transferring technology to peasant farmers are finally receiving the attention they should have had decades ago. My interest in this somewhat esoteric area of study began in Bolivia, where, as a Peace Corps volunteer, I worked as a regional advisor in a community development program sponsored by the U.S. Agency for International Development in several Quechua villages. Among many objectives in the program, the volunteers attempted to improve the dissemination of technical information to small-scale Quechua farmers. The overall program, initiated in four departments of Bolivia, failed miserably.

Although one of the predominant surface reasons for the program's failure was inadequate sporadic funding (the Bolivians did not support their part of a bilateral agreement), the real cause was its lack of a logical organizational arrangement to diffuse knowledge to farmers. Additionally, the Bolivian extension agents who disseminated information were often condescending and showed no real desire or ability to help their clients, the farmers. Even though the farmers strongly desired to receive technical information, they did not trust the extension agents who were supposed to supply them with the information. Finally, even if these two groups could have worked together, Bolivian technological support institutions had few innovative methods relevant to the farmers' needs. Communication techniques are important, but good communication without quality technology accomplishes nothing.

Despite the failure of the program in Bolivia, the two-year experience in that country had a direct effect on the design for two Mexican studies done in 1971, 1972, and 1974. First, it was recognized that an effective

This chapter is based on the author's *Organization for Rural Development: Risk Taking and Appropriate Technology* (New York: Praeger Publishers, 1978).

organizational arrangement for the transfer of technical information to small-scale farmers must provide for the development of trust between recipients and disseminators of technical information. Second, it was realized that the recipients must have some decision-making control that would make participation in a program that undeniably affected their lives worthwhile to them. And third, it was perceived that the arrangement must assist the largest number of recipients at the lowest possible cost—an essential requirement of any transfer program in a developing country.

As an example of a successful technology diffusion program, this chapter will describe the Puebla Project, now called Plan Puebla. Plan Puebla was originally sponsored by the International Center for the Improvement of Wheat and Corn and the Mexican government to increase corn production among subsistence farmers in the state of Puebla. The project is now fully controlled and directed by the Mexican Department of Agriculture.

In accordance with the first criterion explained above, the plan emphasized that to positively benefit the farmer, a sense of trust between extension agents and farmers must be created. To achieve this goal, a rigorous screening process to obtain properly motivated extension agents was used. The agents selected had received instruction not only in agronomy but in rural diffusion techniques incorporating the cultural impact concept derived from cultural anthropology.

In agreement with the second criterion, in Plan Puebla small groups of farmers have always been the recipient units of technical information. This arrangement allows farmer participants as a group to have a decision-making capability in the informational and agricultural input transfer process. The use of many small groups with a small number of extension agents also reduces the cost of the program significantly and satisfies the third criterion. In fact, five extension agents were able to significantly assist 6,000 farmers.

The use of small groups has been the key to success for this technical diffusion project, and a rich and varied literature supports its basic workability. The more relevant references in this case are Abraham H. Maslow's concept of need satisfaction as a motivator for participation, and Rensis Likert's linking-pin integrative organization model.

## Maslow's Concept of Motivation

In his concept of a hierarchy of need satisfaction, Maslow explains that a technical diffusion organization that does not satisfy the social and psychological needs of its members will fail to achieve full realization of its potential.[1] If one extends his ideas to peasant societies, recipients desire more than satisfaction of physical needs by cooperating with an organization. They must satisfy social and perhaps other needs in their collaboration with an information agency. They also need to participate in the decisions on objectives that affect their lives so they will be an integral part of the transfer process. While some may feel that these ideas are an inappropriate adaptation of a concept derived from North American business schools, they nevertheless reveal why Plan Puebla is effective in disseminating technical information to peasants.

If one analyzes the behavior of the Plan Puebla participants, he will note that Maslow's first hierarchial stage, that of physical needs, is easily achieved. By participating in the program, farmers receive technical information, credit, and fertilizer necessary for increased production. Social needs, the second Maslow stage, are also satisfied. Working in groups, participants interact with each other and their extension agent, and they collectively decide and vote on issues that directly affect them. The third Maslow stage, egoistic needs, are satisfied to a more limited extent. Every group elects a group leader who serves as the official link between the group and the higher levels of the transfer organization. These leaders influence other members as well as extension agents, and, in effect, exercise control over other people. Some individuals satisfy self-actualization or creative needs. Farmers are invited to express their opinions on radio, and some villages have musical groups who perform over radio programs directed to participants in the plan.

## The Likert Model

Given this environment of trust, participation, and belief in the credibility of the plan's organization, a strong case can be made that it is consistent with Likert's linking-pin organization model, a further demonstration of its communication and diffusion effectiveness.[2]

The essential concepts of the linking-pin model include:

1. Small group processes maximize the motivation of participants to work in organizations.
2. Group overlapping creates linkages in the organizational hierarchy that help to channel motivation toward group goals and objectives.
3. Group linkages provide a network of feedback loops from higher to lower and vice versa.
4. Representation in decision making through participation in the groups, and established linkages with higher authority, facilitate adoption and change in accordance with the organization's overall objectives.

A visual arrangement of the model (Figure 1) shows a large number of small groups, each with an elected leader forming the base of the organization. The elected leader is, in turn, included in a second-order control group represented by a large triangle intersecting all groups at the elected leader position. In this arrangement, these group leaders can report to the second-order control people and ultimately back to the group. Triangles representing the second-order people are intersected by an even larger triangle representing the top coordinators in the organization.

In Plan Puebla, as stated earlier, decision making on matters that affect participants is based on group processes. Each group elects a leader who formally participates in the second-order control group, the extension agents. The leader serves as liaison or linkage and provides feedback between these two groups. In the final loop, the extension agent group provides feedback to group leaders and group members from the highest control and coordination levels. The primary mode of communication is obviously vertical, yet the organizational arrangement of Plan Puebla has resulted in some unique variations. In a coercive authoritarian structure, the mode of communication would be entirely from the highest to the lowest, with little feedback from the lower levels. In Plan Puebla, at any given time the initiation of communication may come from the lowest, highest, or second level of control. At times the third-level coordinators move to the second level and serve in the same role as extension agents: They inform groups and group leaders of future and present objectives of the program and get their own direct feedback from the lowest level.

An important deviation from the model is that group members are not absolutely required to express themselves through the hierarchy (Figure 2). If they wish, they may and often do communicate directly with the second or third levels.

This formally accepted way to go around the structure provides an escape valve for participants who might otherwise quit the Plan if they could not communicate misgivings about particular objectives without going through men they may not completely trust (either group leaders or extension agents). When such problems arise, they have a formally recognized way to work around them. Many farmers have said that this is an outstanding feature of the organization: knowing that they can go directly to top officials and be listened to with respect gives them an even greater feeling of participation.

A final comment in support of the open communication, linking-pin integration arrangement concerns primarily the second-order control group, the extension agents, who are, after all, the crucial part of the

FIGURE 1

Likert's Linking-Pin Organizational Model with Modifications for Change Agencies Transferring Technology. (Reprinted by permission of the publisher from Rensis Likert, *New Patterns of Management* (New York: McGraw-Hill, 1961).)

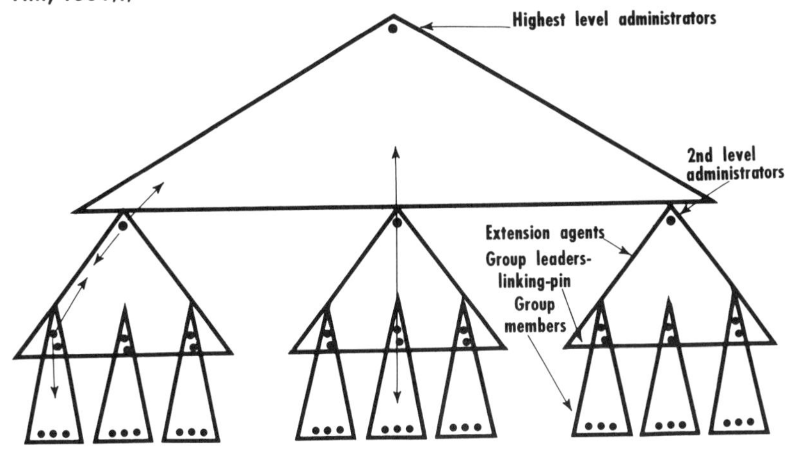


system. If the agents are not effective, group members have no way to obtain the information they need.

Warren G. Bennis, in a related work, states that one should be aware of the extension agents' social and psychological backgrounds, since these will bear directly on their effectiveness as linking-pins between the "cultures" of the small groups and the third-order coordinators.[3] Again, whether by plan or by accident, the plan successfully incorporates these relationships. Most students of the National School of Agriculture at Chapingo, where extension agents receive their training, come from poor family backgrounds. Their education is free, provided they meet the tough intellectual requirements of the school. The product of these circumstances is a bright, well-trained agronomist who understands the problems and frustrations of farmers because he is often one of them. Bennis would argue that such extension agents are effective because they have more in common with their constituencies than with the higher levels of the organization they work for. In the case of Plan Puebla, this commonality of interests is a real reason for their effectiveness with participants.

**Future Linking-Pin Arrangements**

A major problem in Plan Puebla has been the lack of linkage with

**FIGURE 2**
**Formally Sanctioned Ways to Bypass Hierarchical Communications.**

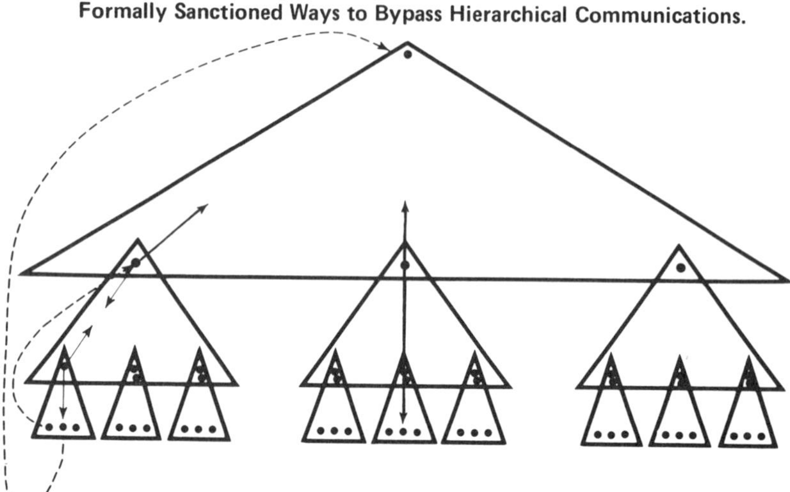

information agencies other than those directly involved with the program. Increasingly, plan participants find they need information from other agencies that they are unable to acquire from their own extension agents. Often, these agents try to obtain the information, but because they have not received training in other areas, they are unable to explain it adequately to the farmers.

Yet, when farmers try to solicit help from other agencies in the State of Puebla, they encounter the same problems in communication common before participation in the plan: that is, their own feelings of inadequacy and little help from agency officials. By expanding the linking-pin model a bit, the plan agents could serve as linkages to these agencies. Such a cooperative arrangement would require the investment of a two-way radio receiving system in the agents' trucks and a clearing office for each agency.

In these other agencies, one can conceive of overlapping triangles of authority without the lowest level of group participants that Plan Puebla had. At the base of the second-order authority, a clearinghouse office with a two-way receiver could be established to help coordinate the activities of the extension agents in both groups (Figure 3).

For example, if a Plan agent were working with a village that needed help on dairy cattle, he could call the clearing office directly via two-way radio, arrange for a meeting between that agency's extension agents and the village group, and possibly have agency-trained people answer direct questions over the radio system. Agents can thus arrange meetings

FIGURE 3
Horizontal Linkage with Other Technology-Transfer Agencies.

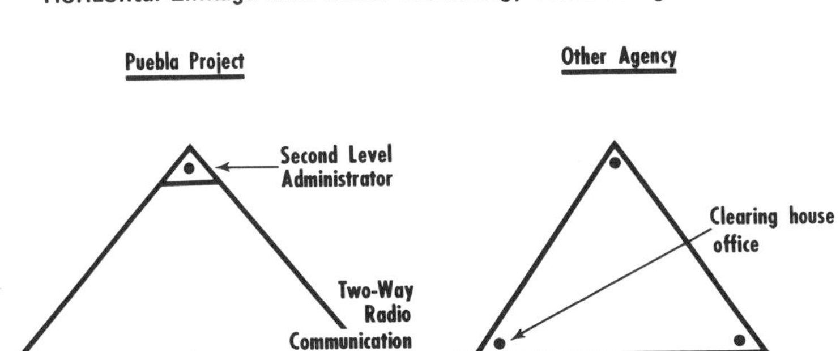


which group participants would hesitate to request on their own initiative. In other words, with an additional small investment, the information-dissemination capabilities of the Plan Puebla agents could be greatly expanded. To complement this approach, it would be beneficial to include in the extension agent's curriculum a course to help him recognize and diagnose problems in the field that are not common to his specific training.

## Developments Elsewhere in Mexico

Discussions with extension agents and farmers indicate the Plan Puebla and its immediate offshoots, Plan Tlaxcala and Plan Zocopiapan, are presently the only effective and innovative technical diffusion programs in Mexico; nevertheless, the immediate future looks brighter for subsistence farmers.

Because of the concern of the Echevarria government to increase the production levels of subsistence farmers, Plan Puebla's strategy and philosophy were used as the model for subsistence farmer improvement programs throughout the country. It was expected that within two years the concepts of small group development and participative decision making by farmers in programs that affect their lives would become as familiar as the cry of *pan y tierra* in earlier years. This approach, and superior variations to come, should make the process of transferring technical knowledge to subsistence farmers in Mexico and other less developed nations far more effective than it has been in the past.

**Notes**

1. Abraham H. Maslow, *Motivation and Personality* (New York: Harper and Row, 1954).

2. Rensis Likert, *New Patterns of Management* (New York: McGraw-Hill, 1961).

3. Warren G. Bennis, Kenneth D. Benne, and Robert Chin, *The Planning of Change* (New York: Holt, Rinehart, and Winston, 1969).

# 8

# Performance and Technology of U.S. and National Firms in Mexico

*Loretta G. Fairchild*

This study was designed to provide a comparison of the behavior of joint ventures and Mexican firms, with special reference to product and process innovation and sources of technology.[1] In addition, at the conceptual level, it adds to the body of empirical data related to "dependency" theory, which suggests that national industry in Latin America has been co-opted by the multinational firm (in part because of their relative disadvantage in the area of science and technology) and consequently is incapable of providing dynamic impetus to the development process in line with national objectives.

Monterrey, Mexico was chosen as the initial research site because it is a heavily industrialized area where domestic firms have maintained a high degree of independence, both from the national government and from multinational firms. In spite of close physical proximity, the question of U.S. influence on the community is very complex. The area is very Mexican in its character and attitudes and is generally less receptive to U.S. investment than is Mexico City.[2] Further information on the composition of the sample and on Mexican policy with respect to foreign investment and industrial development is given in the Appendix.

Intensive interviews with general managers and accounting data for individual firms were used to provide the basis for comparison of

Reprinted, with changes, from an article that appeared in *Journal of Development Studies* 14 (October 1977):14-34.

This study is one part of a multicountry project sponsored by the Program on Policy for Science and Technology in Developing Nations, Cornell University, Ithaca, New York. Similar research was conducted in Brazil, Colombia, and Guatemala during 1975-1976. The author gratefully acknowledges the assistance of Tom E. Davis and Jerry L. Ingles in the development of the research design and analysis.

innovative activity and sources of technology used by the Mexican and joint-venture firms, and of their relative performance, measured in terms of profitability, growth, and export-orientation.

Static indicators used to measure performance for the two groups of firms include return of equity,[3] return on capital,[4] and export orientation (see Table 1). There are no significant differences in level of performance between the two groups. In 1969, return on equity was virtually identical for the two groups. Return on investment appears slightly higher for the joint ventures, but the difference is not significant at the 5 percent level. By 1973, profitability had increased for both groups of firms, but there was still no significant difference.

A great deal of significance should not be attached to the mean estimates of profits. Both domestic and joint-venture firms have an incentive to hide profits. Managers of joint-venture firms in the community generally expected that their declared profits would be higher than those of comparable domestic firms, since as "outsiders" they felt constrained to conform very strictly to the letter of domestic tax laws. To the extent that their expectations were justified, the mean profitability for the domestic firms may be understated relative to the mean profitability for the joint ventures.

Even though it is true for the joint-venture firms that transfer pricing can maximize after-tax profits by redistributing accounting profits within a firm's operation, this does not indicate automatically the direction of such transfers. If an attempt is made to equate the effective marginal tax rate in every area by such a redistribution, for a country with relatively low effective marginal tax rates, the expected direction of such transfers would be in rather than out.[5]

Although no similar information is available for the Monterrey sample, when the study was replicated in Medellín-Cali and São Paulo, only 40 percent of the joint-venture firms in the Colombian sample, and 18 percent in the Brazilian sample, made any purchases from or sales to parent firms that would allow such transfers. Thus, it cannot be assumed a priori that this mechanism is universally available to joint-venture firms.

In order to determine whether joint ventures were "expensing" outlays that domestic firms were capitalizing, cash flow was used as an alternative measure of profitability.[6] Differences are not significant for these ratios either (see Table 1).

A general expectation about joint-venture superiority in exporting was clearly not valid for the Monterrey sample, since the percent of sales exported was the same for both groups as of 1969 and the number of joint-venture firms doing any exporting fell very sharply after 1969,

TABLE 1.[a]

STATIC PERFORMANCE MEASURES

| | 1969 Mexican Firms | 1969 Joint Ventures | 1973 Mexican Firms | 1973 Joint Ventures |
|---|---|---|---|---|
| Profitability: | | | | |
| Return on equity | 11·5% (7·4) | 14·1% (16·5) | 13·3% (13·1)[b] | 17·0% (23·0)[b] |
| Return on investment | 5·8% (3·2) | 7·2% (9·6) | 6·6% (19·6)[b] | 8·4% (9·7)[b] |
| Cash flow/net worth[c] | | | 19·0% (13·7) | 25·9% (14·0) |
| Cash flow/total assets | | | 13·9% (17·7) | 15·3% (10·7) |
| Exports: | | | | |
| Per cent of Sales Exported | 2·8% (9·1) | 5·1% (10·7) | 2·8% (5·8) | 4·0% (11·0) |
| Per cent of Sales Exported[d] | 12·8% (6 firms) | 12·0% (14 firms) | 8·8% (6 firms) | 26·6% (3 firms) |
| Any exports during the period 1969–73 | | | 9 firms | 6 firms |

a. In all tables, data presented are means for each group, followed by standard deviations in parentheses unless otherwise indicated. * indicates significance at the 10% level, ** for significance at the 5% level, and *** for significance at the 1% level, using a two-tailed test.

b. Medians are presented because the means are distorted by a few extraordinary values. Nevertheless, none of these means are significantly different even at the 10% level.

c. One case with an abnormally low net worth in this year, relative to all other years, was omitted from the calculation.

d. Percentages are means only for the number of firms indicated. i.e. only those with response greater than zero.

TABLE 2
STATIC PERFORMANCE: MEANS OF DIFFERENCES FOR PAIRS[a]

| | *1969* | *1973* |
|---|---|---|
| Profitability: | | |
| Return on equity | −0·09% (19·36) | −7·4% (19·93) |
| Return on investment | −1·48% (11·48) | +3·1% (19·71) |
| Cash flow/net worth[b] | | −3·7% (11·1) |
| Cash flow/total assets | | +3·3% (16·8) |
| Exports: | | |
| Percent of annual sales exported | −2·56% (15·58) | +0·06% (10·11) |
| Any exports between 1969 and 1973 (yes) | | +0·71% (3·27)[c] |

a. Differences are calculated as: Mexican performance value minus J.V. performance value for each industry pair.
b. One case with an abnormally low net worth in this year relative to all other years was omitted from the calculation. Differences were not significant at the 10% level even with this case included.
c. Yes–No responses were coded as 5–0 due to an earlier effort to standardise all variables on a 1–5 scale for combining into indices.

while the number of Mexican firms exporting rose.[7]

Another way of presenting the two groups' performance is to pair the joint ventures with "corresponding" domestic firms and measure differences in their performance.[8] None of the mean differences was significant at the 10 percent level (see Table 2). In other words, on a pair-by-pair basis, the Mexican companies were as profitable and exporting as much as their joint-venture competitors.

Firms' performance over time is indicated by average growth of profits, sales, assets, and employment for the two periods 1966 to 1969 and 1969 to 1973 (see Table 3). The same picture is brought out even more clearly by means of differences of growth rates for each pair (see Table 4). In the 1969 period, the joint ventures were growing significantly faster than their Mexican counterparts in sales and total employment. However, by the later period, growth rates tended to equalize, so existing differences were not significant.

On the basis of the 1969 results, it was concluded that the Mexican companies were performing "on par" with the joint ventures, so their survival was not seriously threatened by this foreign competition.[9] It was also predicted that growth patterns over time were tending to mitigate rather than intensify existing differences. The evidence from the second

## TABLE 3

DYNAMIC PERFORMANCE MEASURES

| | *1966–69* | | *1969–73* | |
|---|---|---|---|---|
| | *Mexican Firms* | *Joint Ventures* | *Mexican Firms* | *Joint Ventures* |
| Growth: (annual average)[a] | | | | |
| of Profits | 15·4% (21·0) | 15·4% (22·1) | 15·6% (50·5)[b] | 15·5% (35·1)[b] |
| of Sales | 11·9% (13·0)* | 19·1% (16·6)* | 11·9% (10·2) | 15·0% (7·6) |
| of Assets | 19·7% (18·5) | 15·0% (12·9) | 12·5% (9·8) | 9·5% (11·6) |
| of Total Employment | 6·8% (21·5)[b] | 8·2% (7·6)[b] | 2·6% (9·8)[c] | 6·0% (7·5)[c] |

a. Percentages are calculated as follows: (final year's value minus first year's value) divided by the number of years minus one, to give the average growth rate, during the period, on an annual basis.

b. Medians are presented because the means are distorted by a few extraordinary values; they are not significantly different even at the 10% level, however.

c. Barely significant at the 20% level.

* Indicates significance at the 10% level.

TABLE 4

DYNAMIC PERFORMANCE: MEANS OF DIFFERENCES BETWEEN PAIRS

| | *1966–69* | *1969–73* |
|---|---|---|
| Growth: | | |
| of Profits | −0·21% (20·20) | −1·72% (64·39) |
| of Sales | −8.58% (24·50)* | −1·89% (11·08) |
| of Assets | +2·50% (19·22) | +2·16% (15·32) |
| of Employment | −4·83% (10·62)** | −2·87% (11·51) |

*Indicates significance at 10% level.
**Indicates significance at 5% level or less.

survey, through 1973, would seem to support and reinforce both conclusions.

What factors are enabling the Mexican firms to compete so well? That question was asked of the businessmen when they were reinterviewed in 1974.[10] All but one agreed that the study's conclusion that the Mexican firms were performing equally with the joint ventures was probably accurate. However, the reasons suggested by the two groups were very different.

One half of the general managers in the joint venture firms believed that the Mexican firms were roughly equal chiefly because they relied on foreign technology.[11] The other half suggested that the Mexican companies were making high profits because of questionable management practices—for example, using "outdated" technology and realizing unsustainably high profits because they were not reinvesting. It was also mentioned that the two groups could not really be compared, since their markets were essentially different—only domestic for the Mexican firms and "international" for the joint ventures.

On the other hand, general managers of the Mexican firms suggested that the Mexican companies were doing equally well because of some advantages in being national. (They also cited various disadvantages of having foreign capital.) They stressed the importance of knowing the Mexican market, the fact that foreign technology is transferred with no changes and is badly adapted to local conditions, and that long lags as the firm waits for decisions "from New York" are detrimental. Finally, foreign executives are also seen as primarily concerned with how they appear in their "division," so they focus on dividends and "bleed" the Mexican operation, jeopardizing its long-run growth.

Opinions among both groups were split evenly on whether or not

Mexican companies have any advantage in terms of consumer preference or getting credit, but they were considered to have an advantage in any dealings with the government, such as import permits. On the question of whether foreign firms really try less hard because they feel vulnerable, opinions were evenly and sharply divided in both groups.

An examination of the empirical data provides some insight on the validity of these opinions and at least suggests why the Mexican firms are performing so well.

One variable which, according to development literature, should provide significant competitive advantage to the joint ventures, is access to technology that results in process and product innovation. The data fail to show that the joint ventures are significantly more innovative than the national firms (see Table 5).[12] Instead, while performance was very nearly equal, the slightly significant differences that do exist indicate greater Mexican innovation.

With respect to product changes, by the 1969–1973 period, more Mexican firms added products at increasing levels of sophistication. For the joint ventures, roughly 20 percent made no changes in either period, and the percentage making the most complex changes decreased.[13] Mexican superiority in the second period in the complexity of products added is almost significant at the 10 percent level.

Although roughly the same percentage of both groups indicated that some changes in production processes had been initiated, the size of changes was nearly one-third higher for the Mexican firms in the later period, and the difference seems to have been increasing over time.[14]

This high level of innovation by the Mexican firms might be the cause or the consequence of the fact that they are relying heavily on imported foreign technology, brought in under technical assistance contracts or from foreign technicians. However, this frequently heard hypothesis is not confirmed by the data.

Between the two periods, as innovation by Mexican companies increased, their use of patents and technical assistance contracts was low and did not increase (see Table 6). In the second period, twice as many joint ventures had technical assistance contracts,[15] and *four* times as many were using U.S. patents and/or licenses.[16] Only twelve of the twenty-five Mexican companies had any kind of formal contract with the United States or any other country, and for three of them the agreements had expired some time during the period. Thus, for only nine companies (36 percent) were these contracts operational as of 1974.

Estimates on royalties show this same trend.[17] Few Mexican firms paid any, and the number that did decreased over time.

TABLE 5

INNOVATION

| | *Mexican Firms* | *Joint Ventures* | *Mean Differences* |
|---|---|---|---|
| 1969 Survey | | | |
| 1. Most complex change in product | | | −0·28 (1·72)[a] |
| model variation | 27% | 49% | |
| new lines | 40% | 20% | |
| new types | 3% | 14% | |
| any changes (yes) | 70% | 83% | |
| 2. Per cent of new processes[b] | 20% | 21% | +0·31 (2·48) |
| 3. Any changes in processes, within 3 years (yes) | 70% | 49% | +1·0 (3·57)[c] |
| 4. Products dropped in last 5 years (yes) | 46% | 28% | |
| 1973 Survey | | | |
| 1. Most complex product change made | | | 0·17 (0·49)[d] |
| model variations | 36% | 30% | |
| new lines[e] | 56% | 44% | 0·65 (3·47) |
| new types[e] | 8% | 4% | 0·22 (1·83) |
| any changes (yes) | 100% | 78% | |
| 2. Per cent of new processes[e] | 28% | 20% | 9·09 (32·76)[f] |
| 3. Any change in processes, within 3 years | 68% | 70% | 0·0 (2·61) |

a. All figures in parentheses are standard deviations.

b. 'New' is introduced within the last six years.

c. The average difference here is positive, indicating typical Mexican changes, while their J.V. partners had introduced none. The mean of the difference is significantly different from zero at the 15% level.

d. Significant at 15% and very nearly significant at the 10% level.

e. 'New' is introduced within the last three years.

f. Almost significant at the 20% level.

TABLE 6

USE OF FORMAL INSTRUMENTS FOR TRANSFERRING TECHNOLOGY

| | *Mexican Firms* | *Joint Ventures* | *Mean Differences* |
|---|---|---|---|
| 1969 Survey | | | |
| Number of firms using: | | | |
| 1. U.S. T.A. contracts | 7 | 20 | −2·14 (2·86)** |
| Foreign T.A. contracts[a] | 1 | 1 | −0·0 (1·36) |
| 2. Any U.S. patents | 5 | 16 | −1·90 (3·11)** |
| Any foreign patents | 0 | 2 | −0·34 (1·29)[c] |
| 1973 Survey | | | |
| Number of firms using: | | | |
| 1. U.S. T.A. contracts | 8 | 16 | −0·32 (0·95)[c] |
| Foreign T.A. contracts | 3 | 4 | −0·09 (0·67) |
| 2. U.S. patents | 3 | 11 | 1·95 (13·04) |
| Foreign patents | 2 | 2 | −1·00 (4·47) |
| 3. U.S. licenses | 2 | 9 | −0·14 (0·48)[c] |
| Foreign licenses | 1 | 4 | −0·09 (0·43) |
| 4. Average royalties paid:[b] 1969 | 54 (7 firms) | 63·8 (20 firms) | −46·9 (92·1)** |
| Average royalties paid:[b] 1973 | 105·6 (5 firms) | 109·4 (14 firms) | |
| 5. Average of royalties/net sales: 1969 | 0·50% (·011)** | 1·97% (·027)** | −0·03 (0·03)** |
| Average of royalties/net sales: 1973 | 0·45% (·011)** | 2·06% (·029)** | −0·02 (0·03)* |

T.A. refers to 'Technical Assistance'.

a. From any foreign country other than the U.S.

b. Averages, in 1,000's of U.S. dollars, for the number of firms shown, i.e. those paying any royalties.

c. Indicates significance at 20% level.

* Indicates significance at 10% level.

** Indicates significance at 5% level.

Foreign technical information may also be imported by employing foreigners. However, Mexican firms throughout both periods seem to have relied very little on foreign engineers, either hired directly or brought in as consultants (see Table 7).

Use of U.S. engineering consultants since 1966 has been consistently higher for the joint ventures.[18] Only 35 percent of the Mexican firms used any foreign consultants in the first period and 24 percent in the second. While their use of U.S. consultants dropped sharply, use of non-U.S. foreign consultants rose some. Such a shift within the pattern of declining use is also important, since greater variety in technical sources would lessen the degree of dependence. Only two Mexican firms had hired any U.S. engineers or administrators in the first period, and by 1973 none were using any.[19]

The empirical evidence seems to indicate that Mexican firms are competing well by relying, not on foreign technology brought in through formal channels, but rather on local consultants, often drawn from the universities, and on the firms' own internal technical capabilities.

In contrast to the relatively low use of foreign technical consultants, Mexican companies have relied more heavily on Mexican consultants. Perhaps their importance is underlined by the extraordinary jump, between the two periods, in the use of Mexican consultants by the joint ventures. In 1969, 31 percent of the Mexican firms utilized Mexican engineering consultants, while 40 percent did so by 1973; joint venture usage rose from 9 percent to 30 percent. In spite of this increased use by joint ventures, however, Mexican company usage is still significantly greater in both periods.

Further evidence on the importance of internal sources of information is brought out in another way by the opinions of the general managers on what they considered to be the most important sources of technical help available to them as they tried to solve specific problems.

For the original technical information used as the firm was founded, approximately 70 percent of the Mexican companies had relied on the founders or other local technicians, and only 30 percent had used foreign companies or technicians.[20] For the joint ventures, however, 57 percent had received the information from the U.S. partner, and 80 percent had used U.S. companies or technicians.

For the information used in achieving process innovation in the earlier period, 75 percent of the Mexican firms said their own administrators or engineers were the chief source for technical information, and only 16 percent relied chiefly on foreign sources.[21] The joint ventures were evenly split, with 40 percent relying on their U.S. partner and 40 percent using ideas from their local administrators and engineers.

TABLE 7

USE OF FOREIGN CONSULTANTS AND EMPLOYEES

| | *Mexican Firms* | *Joint Ventures* | *Mean Differences* |
|---|---|---|---|
| 1969 Survey: | | | |
| Use of U.S. engineering consultants | 35% | 47% | −0·83 (3·81) |
| Use of foreign engineering consultants | 0% | 6% | −0·21 (1·02) |
| Any U.S. engineers and/or administrators employed directly | 2 firms | 15 firms | −0·94 (1·8)* |
| Per cent of U.S. engineers and/or administrators[a] | 8%[b] | 18%[b] | −0·10 (0·17)** |
| 1973 Survey: | | | |
| Use of U.S. engineering consultants | 16% | 44% | −1·25 (2·75)** |
| Use of foreign engineering consultants | 8% | 4% | +0·25 (1·12) |
| Any U.S. engineers and/or administrators employed directly | 0 | 10 firms | |
| Per cent of U.S. engineers and/or administrators[c] | — | 18% | |

a. Calculated as a percent of the total number of engineers plus administrators in the firm.
b. Average for only those firms having any foreign engineers or administrators (10% of Mexican firms and 58% of J.V. firms).
c. Average for those having any U.S. personnel, as a percentage of the total number of engineers.
* Indicates significance at 10% level.
** Indicates significance at 5% level or above.

Over time, there seems to have been little change among the Mexican firms; they are still relying most heavily on their own "idea people." Among the joint ventures, however, there seems to be a slight decrease in reliance on the parent; an increasing percentage are centering R&D efforts within the firm and not relying exclusively on foreign R&D.

In addition, there are more objective measures of internal technically oriented activity that make this reliance on internal sources possible (see Table 8). Putting formal attention on developing new processes[22] seems to be gaining momentum over time, since relatively few firms were active in this area before 1971, while well over half have begun to do so since that time.[23]

Tangible results have come from this research activity. Mexican firms were developing and registering their own patents significantly more often than their competitors prior to 1969, and the basic trend seems to be continuing. It should also be noted that patents do not reflect all the activity in this area. Several firms with "patentable ideas" explained that they had not obtained patents simply because they were afraid this might allow the information to reach the "wrong hands." Or, since the machinery was not for resale, they often felt it was not worth the bother to obtain a patent.

One-third of the Mexican firms had designed and built some of their own machinery—significantly more than was done by the joint ventures—in contrast to assumptions generally made that LDCs are not producing any of their own technology. Research and development expenditures are still relatively small scale, but their very existence, and the fact that more and more Mexican firms are spending on R&D, would seem to have important implications for Mexico's technological future.

In the light of this evidence of relatively more internal R&D activity among Mexican firms, it should be noted that managers of the Mexican firms perceived much greater obstacles for their firms in obtaining needed technology. Very few joint ventures felt that there were any obstacles, even though the Mexican firms were introducing the most changes. Managers were asked to rank in importance the obstacles they had faced in obtaining technology. The two most common responses were (1) that the level of automation was too high for the Mexican market, and (2) that the technology was available but too costly.

Difficulty in obtaining technical information seemed to vary substantially by industry, chiefly between those perceived as "open" rather than "closed." When very specialized technology is used, only people within the industry are likely to be knowledgeable about the latest developments. If an industry is "open," if competitors talk to each other, much more information seems to be available for fomenting new

## TABLE 8
INTERNAL R & D ACTIVITY

| | *Mexican Firms* | *Joint Ventures* | *Mean Differences* |
|---|---|---|---|
| 1. Some R & D Effort on processes: | | | |
| before 1971 | 28% | 26% | 0·22 (2·19) |
| beginning 1971 | 68% | 52% | 1·09 (3·67)[a] |
| 2. Formal efforts on quality control | 64% | 93% | −1·59 (2·38)** |
| 3. Own patents registered[b] | | | |
| before 1969 | 5 firms | 3 firms | 0·91 (2·02)* |
| beginning 1969 | 6 firms | 4 firms | 0·43 (2·5) |
| 4. Designed/Built any of machinery used | 33% | 15% | 1·14 (3·06)* |
| 5. Average amount spent on R & D:[c] 1969 | 44·0 (10 firms) | 22·2 (6 firms) | 16·2 (41·9)[d] |
| Average amount spent on R & D:[c] 1973 | 40·5 (14 firms) | 51·0 (8 firms) | |
| 6. Average of R & D expenditures/net sales: 1969 | 0·30% (·005) | 0·33% (·010) | |
| Average of R & D expenditures/net sales: 1973 | 0·55% (·007)[d] | 0·21% (·003)[d] | 0·003 (0·008) |

a. Indicates significance at the 20% level.

b. Firms gave the number of patents they or their engineers had developed and registered in each period. The table indicates the number of firms registering any. The number of patents registered by Mexican firms dropped for the second period. However, this is probably due to the fact that five years is simply too short a period to measure such activity, which traditionally has a long gestation period. Patents developed elsewhere which the firm re-registered in Mexico were explicitly excluded.

c. In thousands of U.S. dollars. Averages are for the number of firms given in parentheses, i.e. those doing any R & D spending. Figures given were 'best estimates', not exact data.

** Indicates significance at the 5% level, or better.

* Indicates significance at the 10% level.

d. Indicates significance at the 15% level.

ideas. Perhaps industry behavior with respect to communication channels is a variable which should be explored more explicitly in the future.

In considering these attitudes toward the search for technical information, it is interesting to speculate that perhaps since the joint-venture firms have an established "channel" (i.e., from the U.S. parent company), the entire issue of obtaining technology creates less anxiety, since less search is needed and, therefore, less activity is undertaken. The basic position of the subsidiary might be considered passive.

For the Mexican firms, on the other hand, the problem looms larger. They are more "anxious" and uncertain about what can and should be done, but they are therefore more active—which shows up in the performance and innovation variables.

In conclusion, the conception of the Monterrey industrial community suggested by these data differs significantly from that contained in a large part of the "dependency" literature. Not only are established Mexican firms in a broad range of industries performing comparably with foreign firms in terms of profitability, growth, and exports, but they appear to be at least as innovative in the sense of introducing new products and productive processes. Perhaps even more important, they appear to be relying substantially on domestic institutions and particularly on resources internal to the firm to generate the new technology. To say the least, a national industrial base not subordinated to foreign competition appears to be alive and well and living in Monterrey.

## Appendix: Composition of the Sample

The initial interviews were conducted during the period 1969 to 1971. These same manufacturing firms were reinterviewed during June and July of 1974. Manufacturing firms are included from the following five product classifications: (1) metal products; (2) nonmetallic minerals such as glass, bricks, etc.; (3) chemicals; (4) food, beverages, and tobacco; and (5) assembly. Each one of the joint-venture firms had some U.S. direct equity investment. Each one was then matched with a firm that had 100 percent Mexican capital and made basically the same product. Wherever possible, the chief competitor was chosen. Other matching criteria were size, measured in terms of net sales, and age. Only firms in operation more than three years were included, since it was felt that the behavior patterns of newer firms were too erratic to indicate general or average performance. (See Table A.1.)

The smallest firm included in the initial sample had annual net sales

TABLE A-1 CHARACTERISTICS OF THIS SAMPLE

| | *1969* | | *1973* | |
|---|---|---|---|---|
| | *Mexican Firms* | *Joint Ventures* | *Mexican Firms* | *Joint Ventures* |
| Age, as of 1969, in years | 18 | 13 | | |
| Size: | | | | |
| Total Assets [a] | 1·1 | 1·7 | 2·7 | 2·6 |
| Net Sales[a] | 1·8 | 2·3 | 4·2 | 3·9 |
| Total Employment | 180 | 160 | 257 | 236 |

[a]Millions of U.S. Dollars

of 80 thousand U.S. dollars, while the largest had net sales of over 200 million U.S. dollars. As of 1973, both groups on average had virtually identical total assets. Net sales were slightly larger for the Mexican firms, but the difference is not significant. Total employment was also essentially the same.

The period 1969 to 1973 was characterized by tremendous growth for Mexican industry in general, although they were very difficult years. Mexico suffered a fairly severe recession during 1970 and 1971. Inflation was worse than in the United States throughout the period, and recently, manufacturers faced a severe scaricty of raw materials.

Mexican policy on industrialization is planned and administered by the federal government. Business managers generally admit that over the last twenty years, business practices have gradually come under much closer governmental scrutiny. Today, federal regulations and reporting policies must be adhered to more closely than managers often find convenient.

Mexican policies to promote industrial development are quite representative of those being used in other relatively advanced industrial areas in Latin America. Import licensing and a complex tariff structure are used to manipulate the direction of industrial development, with a primary emphasis on increasing the degree of export promotion and "integration," i.e., the percent of processing of value-added that is generated in Mexico.

Tax incentives provided to "new and necessary" industries involve considerable administrative costs, making them unattractive to all but larger firms. More use is made of "Rule 14" allowing reduced import taxes on imported machinery for use in manufacturing. Firms complain, however, that its applicability is quite restricted, since the reduced rates apply only on large machinery complexes, not to individual items or replacement parts. Depreciation policies are not generally used as a tool for industrial promotion.[24]

Foreign investment is welcomed but subjected to regulation. It is forbidden in key areas such as banking and finance, transportation and communication, and national defense. In many other areas, foreign investment is permitted only with majority domestic ownership. This category is being gradually expanded as part of a wider effort to increase the degree of Mexicanization. With respect to taxes, import licensing, and other legalities, regulations are the same for foreign and domestic firms.[25]

Once a firm demonstrates to the government that it can produce an item, permits for importing similar foreign items are denied, providing a substantial degree of protection to all producers inside the tariff wall. However, the general incentives and protection provided are not atypical for Latin America.[26]

Although Monterrey is not "typical" of the general business environment in all of Latin America, it is perhaps "representative" of the direction in which industrial development may move. Regional industrialization programs in other parts of Mexico and the rest of Latin America are helping to create similar environments. Preliminary results from a replication of this study in São Paulo and Medellín/Cali show generally similiar "success stories" on the part of domestic firms.

## Notes

1. The sample includes twenty-five manufacturing firms in Monterrey, Mexico with direct U.S. equity investment and a matching firm for each with 100 percent Mexican capital. The firms were interviewed by Loretta Good Fairchild in 1969 and again in 1974. Further information on the makeup of the sample is found in the Appendix.

2. Loretta L. Good, *United States Joint Ventures and National Manufacturing Firms in Monterrey, Mexico: Comparative Styles of Management* (Ithaca, New York: Cornell University Latin American Studies Program Dissertation Series, 1972), p. 23.

3. Return on equity was calculated as before-tax net profit divided by net worth.

4. Return on investment was calculated as before-tax net profit divided by total assets. Using fixed assets as the denominator yielded similar results.

5. J. L. Ingles and L. G. Fairchild, "Empirical Evidence on Industrial Transfer of Technology: Measuring the Impact of Foreign

Ownership Structure in Brazil, Colombia, and Mexico," (Report from Program on Policies for Science and Technology in Developing Nations, Cornell University, Ithaca, N.Y., 1976), p. 19.

6. Cash flow was calculated as annual reported before-tax net profits, plus annual depreciation charges. This figure was then calculated as a percentage with respect to net worth and total assets for comparability with other profit figures.

7. Several exporters among the Mexican firms indicated that their percentage of sales exported was lower in 1973 because of greatly increased domestic sales, not because of any decrease in sales abroad.

8. Differences for each variable were calculated as Mexican performance value minus that of its joint-venture counterpart. A Student-t, two-tailed test was used to determine whether or not the average of the differences was significantly different from zero. Because of the direction of subtraction, the average will be negative when the joint venture did better than the Mexican firm and positive when the Mexican firm was superior. See the Appendix for further information on criteria for matching firms and sample characteristics.

9. Good, *United States Joint Ventures,* pp. 82, 134.

10. Opinions are available only from eight Mexican firms and six joint-venture firms.

11. Of the joint-venture firms expressing opinions, 50 percent were Mexican and 50 percent were American.

12. Innovation is used here to refer to changes introduced, not to R&D activity that might or might not lead to tangible results.

13. "Type" refers to an area of products, all of which are made of the same raw materials. A "line" is a subcategory of products within a type which all have the same specific function. (It is analogous to general names of products: motors, pumps, enamel paint, etc.) If different raw materials are used, but the processing is the same, it would still be considered a single line. "Model" variations are changes only in designs, colors, styles, or sizes. Since the categories were ranked in order of increasing complexity, the difference in "complexity of product introduced" indicated that one firm is equal with its competitor, or "one step," or "two steps" ahead.

14. The general manager was asked to estimate the percentage of processes currently in use that were not in use three years earlier to get an indication of recent innovative activity. Such "guesstimates" will vary in validity among industries. However, pairing of the firms within product classifications minimizes these difficulties. Further analysis has indicated that differences in percent of new processes added are significantly related to relatively successful performance on growth

variables during the second period. Since it was often difficult to get a precise estimate of the change, a related question (Were any significant changes in production introduced in the last three years?—yes/no) was also used. Firms giving specific percentage responses are also included in this latter category. The question on products dropped recently was omitted from the 1973 questionnaire in the interest of brevity. Percentages given indicate those firms out of the total giving a "yes" response.

15. Data were tabulated as the yes/no response to whether or not the firm had utilized any technical assistance contract, etc., any time during the period. Figures were also obtained on the number of technical assistance contracts, patents, and licenses utilized by a particular firm. This information may be overlapping in the sense that a particular license may contain an agreement on conveying technology and several patents. At the other extreme, a firm may hold several patents, each one of which is from a separate company and is covered by a separate contract. The adjective "foreign" is used to indicate contracts with all other countries except the United States.

16. For the joint ventures, there is a fairly strong tendency for the use of technical assistance contracts to be higher for firms with lower percentages of U.S. equity. This is not always the case, of course; for one firm, the percentage of U.S. equity fell from over 80 percent to 49 percent without any change in the basic management structure. Even at 49 percent no technical assistance contract was used and no royalties were charged, even though the firm had originally been allowed 80 percent of the equity as part of an agreement for supplying the technical assistance without charge. Use of brand names is included with licenses.

17. In order to gain a clearer understanding of the order of magnitude of royalties being paid, the averages presented are only those for firms paying any. Figures of royalties to net sales are especially interesting, since similar measures are being used by the Registry of Technical Assistance Contracts as it evaluates the appropriateness of cost to technology received.

18. Visits from people on the staff of the parent company are *not* included, as inputs from owners are viewed throughout the study as "internal to the firm." They may or may not be from the same companies with whom a firm has a technical assistance contract. "Foreign" refers to all other countries except the United States.

19. This second category refers to those on the company's payroll who were brought in specifically because of their expertise. Cuban refugees now living in Mexico, for example, are not included.

20. In the data presented from the 1969 survey, the "chief source of

technical information" refers to that utilized when the company was founded, regardless of the ownership structure at that time. For the 1973 survey, however, the same item refers to the chief source the firm used for technical information since 1969.

21. Firms were asked to rank in order of importance the three most important sources of ideas used in the changes in processes. Similar information for new products proved less useful because the administrator tended to see himself as the chief source, regardless of where he encountered the idea.

22. Firms responded with simply a yes or no as to whether or not any formal attention was placed on developing new processes in each time period. "Formal" indicates that it had become part of an employee's job description, even if he did not work on it full time.

23. The one area in which joint ventures appear to be putting more emphasis than the nationals is quality control. To qualify for a "yes" indicating formal attention on quality control, the firm had to have at least one person working on it full time. Further work is needed in this area to delineate trends and implications.

24. José Miguel Díaz Noriega, "Brief Summary of Mexican Taxation Policy and Practice Based on 1970 Law" (Monterrey, Mexico, 1970), pp. 4-8.

25. Harry K. Wright, *Foreign Enterprise in Mexico* (Chapel Hill: University of North Carolina Press, 1971), pp. 105, 109, 149.

26. "Latin America's Industrial Incentives," *Industrial Development* (January 1967):75.

# 9 Organizational Structure and Innovativeness in the Pulp and Paper Industry of Mexico

*Viviane B. de Márquez*

The impact of transnational corporations on host developing countries has generally been viewed in terms of their global economic effect on the balance of payments, employment, income distribution, or the structure of local capital markets.[1] More recently, the role of the multinational corporation in the transfer of technology and its impact on technical advance in developing countries have become important themes. It has been argued, for example, that multinational corporations transfer mostly routinized technological processes at the "mature" stage of their product cycle or that the technology transferred is usually inappropriate to factor endowments in these countries.[2] Some have asserted that although multinational firms have contributed positively to the development effort of many countries, they have also created and reinforced patterns of technological dependency in these countries both through their monopoly over technical knowledge and through their sheer economic power. In this line of reasoning, it is assumed that domestic sources of technological innovation in the developing countries are nonexistent.

Although the majority of published work generally supports these views, a few dissenting voices do not counter such arguments as much as some of their premises. In the first place, many of the assertions found in the literature apply to fully owned subsidiaries that are slowly disappearing as a result of defensive policies of host governments, to be replaced by joint ventures. Second, the relative success of multinational corporations in the developing countries is not due to the special magic of their superior managerial skills and financial power, but rather to the

The author wishes to thank those whose cooperation has made this project possible, in particular Laura González Durán for her collaboration in the research, Ing. M. Frank and Lic. H. Escoto of the Cámara Nacional de las Industrias de la Celulosa y del Papel, and the firms that participated in the study.

weaknesses of most of the domestic firms with which they are usually compared. Thus, when we find that multinational firms dominate certain sectors, it simply means that few equally large national firms can be found in those particular sectors.[3] On the other hand, when organizational characteristics between national and foreign firms are matched, the latter do not show any systematic tendencies to be more successful or to dominate the market.[4]

Finally, it can be argued that we know too little about the individual behavior of firms in the developing countries (whether foreign or national) to be able to assert that local sources of technological innovation are nonexistent. The concept of innovation cannot be reduced to the number of patents generated.[5] Therefore, this concept must be broadened to include marginal adaptive processes, and their incidence in the developing countries must be investigated.

The study presented here is part of a larger ongoing project of research into technological innovation in two sectors of Mexican manufacturing industry. This chapter reports preliminary results obtained from the pulp and paper manufacturing sector. Its purpose is to investigate differences between national and multinational firms in technological innovation, where the term includes adaptive as well as inventive activities. Organizational characteristics are considered important intervening variables that help explain technological behavior. Because of the small number of cases, results can only be regarded as suggestive.

## The Role of Transnational Corporations in Developing Countries

The first thing that can be said about the literature on transnational firms in Mexico is that it disregards sectorial differences. Yet an important distinction must be made between transnational firms that produce for the export market and those that produce for the internal market of the host country. In the first case, their actions respond to a global strategy oriented toward the world market; in the second, their interests are at least compatible with those of the host country.[6] This study concerns an industry that is entirely oriented toward the Mexican market.

A second observation that applies to the literature on transnational firms in Mexico is that it is inconclusive. Because few conclusions are derived from systematic empirical research based on a wide variety of cases, we find many contradictory assertions. On one point all studies of transnational firms in Mexico agree, however: that they occupy a leading position in the economy in their size relative to national enterprises, their rate of growth, and their location in the more dynamic sectors of the economy.

The ability and willingness of the Mexican subsidiaries of transnational corporations to be innovative technologically or to adapt foreign technology to local needs are much in debate.[7] A study of pharmaceutical, automotive, and petrochemical industries indicates that transnational subsidiaries in Mexico do not, generally, adapt their technology to local conditions except when competing with other transnational firms, in which case they use scaling down techniques devised by their own headquarters (whom they assert make all technological decisions for them).[8] On the other hand, Manfred Nitsch has claimed that transnational firms are more innovative than state-owned firms, but he presents no concrete cases to support his claim.[9] Sepúlveda and Chumacero go further than Wionczek, Bueno, and Navarrete in claiming that whatever technological progress is achieved by local subsidiaries of transnational corporations is appropriated by their headquarters, so the subsidiaries have no incentive to spend resources on research and development.[10] This assertion, however, seems to be based on a few isolated cases, judging from the results obtained by María y Campos in his analysis of the Registro Nacional de Transferencia de Tecnología, in which fewer than 3 percent of the cases studied presented any restrictive clauses either on export policy or on local technological developments.[11]

Transnational firms are also said to be more capital-intensive and, as a result, to create less employment.[12] Fajnzylber finds that they are indeed more capital-intensive, but that they create more employment than other types of firms because of their much higher rate of growth.[13]

Another moot point is the degree of modernization of the equipment used by transnational firms as compared with local firms. A study published by Nacional Financiera shows a higher degree of equipment modernization by transnational firms.[14] Yet Sepúlveda and Chumacero indicate that they use obsolete equipment. The confusion may be due to the failure to specify what equipment in transnational firms is considered obsolete. These firms may use equipment that is obsolete in their country of origin, yet more modern than the equipment of local firms in the host country. In any case, economists do not agree which is better for underdeveloped countries: relatively obsolete equipment, usually smaller and less capital-intensive than newer developments, or the latest equipment that often saves money on both capital and labor. Many case studies in different sectors will be needed to decide this issue.

Structural differences between transnational firms and local firms in Mexico have also been described, but only in terms of economic salaries, all of which are reputed to be higher in transnational firms, according to Fajnzylber and Tarragó. What is lacking is information on the organizational differences that distinguish transnational from local firms. Although scarcity of human resources and organizational deficiencies are known to be crucial to underdevelopment, there are no data on the impact of these factors on the differential rate of growth and technological advancement of transnational versus local firms.[15] Fajnzylber's study reveals that transnational firms tend to monopolize the best skills in their respective sectors, because they offer higher salaries and generally better conditions for the advancement of individual careers. We may also predict that transnational firms will have a closer administrative control of the productive process and be more oriented toward cost-cutting techniques because of their organizational experience in developed nations, where such factors mean the difference between success and failure.

## Sectorial Characteristics of the Pulp and Paper Industry

The manufacture of cellulose and paper in Mexico has been an expanding industry whose products have been intended almost exclusively for the internal market.[16] Between 1966 and 1974, the production of cellulose in Mexico grew at an average annual rate of 8.1 percent, compared with 10.2 percent for Latin America as a whole, while Mexican paper manufacturing had an average annual growth rate of 12.2 percent, as against 10.2 percent for the whole of Latin America. The Mexican paper industry has therefore been increasingly dependent on imports of cellulose, in spite of the steady increase in pulp making.[17]

Cellulose is manufactured in Mexico by the traditional coniferous wood process and by utilizing sugarcane bagasse. For the latter, there are basically two processes, one developed by W. R. Grace and Company in 1939, and the other developed in the 1950s by Dr. Dante Sandro Cusi of the San Cristóbal Company in Mexico. Several processes represent slight departures from the Grace method (with the result that several firms claim to have "developed" the sugarcane bagasse process), and one uses linseed straw as raw material for making cigarette paper.[18] Also significant is the high rate of paper recycling in Mexico: over 40 percent of the raw material used to make paper is waste paper.[19]

In 1976, fifty-six companies were registered with the Cámara Nacional de las Industrias de la Celulosa y del Papel, of which thirteen manufactured paper and pulp, eleven pulp only, and thirty-two paper only. ("Paper" includes the gamut from fine Bible paper to hard cardboard.) The industry is, in fact, much smaller than these figures suggest. It is composed of several groups of companies, most of which operate under the centralized leadership of one company (the top executives of the latter serving as the center of decisions for the others). We found nine such groups, not including *maquiladoras.*[20]

The groups of firms rarely represent cases of vertical integration. They are usually firms that have been partly or fully bought by larger, more successful firms and that keep manufacturing the same product-mix they had before acquisition, even if it appears to be in competition with the parent company.[21] An exception to this rule is cardboard; the manufacture of cardboard is often integrated with the production of goods that need crating.

In addition to being highly concentrated financially, the Mexican pulp and paper industry is also highly concentrated geographically. Of a total of sixty-three plants in 1975, forty were located in the Federal District, the State of Mexico (mainly in the suburbs of Mexico City), or the State of Puebla—that is, in the center of the country. The other three states with a relatively high concentration of plants are Nuevo León in the North with four plants, Veracruz on the Gulf of Mexico with four, and Jalisco on the Pacific with five. The remaining nine plants are widely dispersed throughout the country. Nevertheless, the government's policy of decentralizing industry is beginning to bear fruit; all new plants located since 1965 are found outside the more densely industrialized areas.

As in most Mexican industrial fields, production in the pulp and paper industry is economically concentrated: the four largest companies account for 31 percent of the total volume of production.[22] The degree of actual competition between companies is tempered by the fact that the market is divided among a number of submarkets that are then distributed among the existing companies. The smaller marginal companies tend to specialize, for example, in low-grade wrapping paper for the retail trade, for which little virgin fiber is used; the larger firms orient themselves toward the highly competitive consumer market.

## Structure of the Study

Twenty-two of fifty-six existing firms were investigated between January and October of 1976, although only twenty firms were ultimately retained in the study. The sample obtained is a purposive one, designed to maximize variance on three variables: the structure of ownership (transnational, state, and privately owned), the size of the firm, and regional dispersion. The first variable was defined on an operational basis; any firm that had 30 percent or more foreign capital was considered transnational, and national state-owned firms were defined as those having at least 51 percent of state capital.[23] The other firms were considered as private national. The size of firms was measured by production, using 1975 data. The regional distribution was divided between Center, North, East, and West. The resulting distribution of cases included eight large, eight medium, and four small firms; three transnational and seventeen national firms; and eleven firms located in the Center, four in the North, four in the West, and one in the East.

Among the twenty companies investigated, a total of twenty-seven plants were included. In the results that follow, the unit of analysis will be alternatively the plant or the firm, depending on the problem involved. In the organizational literature, the issue of the relevant unit of analysis is still an open one and has not been discussed extensively. The failure to find any substantial associations between technological and organizational characteristics may arise from taking plants as units of analysis to test that association when they are not centers of power but mere executors of main-office directives. That should be all the more true in a relatively routine industry like the pulp and paper industry, where, in Perrow's terms, exceptions are frequent but problems are easily analyzable.[24] In this report, technological characteristics are based on plant data, whereas technological decisions and administrative structure are described for entire firms.

Respondents in all firms were the highest executives, heads of departments and their superiors. Except for questions dealing with centralization that have attitudinal dimensions, no effort was made to ask questions in a standardized way or in a given order. In the smallest companies, the head of the firm and the engineer in charge of production were usually the only persons interviewed.

## Results of the Investigation

Three sets of characteristics were investigated. The first two may be considered as independent variables and the last, the dependent variable. The first set will be referred to as the organizational configuration. It includes a series of structural organizational characteristics of the firms studied which are indicative of the ways in which information, decisions, and ideas circulate in an organization. The second set includes some basic technological data that describe the productive process. The third set concerns innovative behavior.

### *The Organizational Configuration*

For this part of the analysis, the variables measured for an organization were size, rate of growth, internal complexity, professionalization, centralization, and formalization of procedures. Size was measured by production in metric tons of output. The rate of growth was measured by percentage growth in production between 1970 and 1975. Complexity was measured by the number of major divisions at the firm level and by the number of professional specialties represented in the first three levels of hierarchy. Centralization of decisionmaking was divided into two aspects: routine decisions and general policy decisions. Heads of departments were asked if they felt they participated in general policy decisions and if heads of firms interfered in the daily routine of their departments. Concurrently, heads of firms were asked if they allowed their heads of departments to participate in either kind of decision making. In addition, the frequency of formal meetings between heads of firms and heads of departments was taken as an indicator of centralization, on the assumption that a relatively routine type of production technology such as that in the paper industry requires a relatively low frequency of scheduled meetings.[25] A fourth item on centralization was a question asked of first-level executives, whether the head of the firm "exercised strong leadership."

Formalization was measured by the presence or absence of a manual of procedures and by the degree of personnel evaluation. Professionalization, on the other hand, was measured as the percentage of management personnel with a B. A. equivalent or an engineering degree.

As Table 1 shows, two of the three transnational companies surveyed are large firms. It also shows that they do not dominate the sector, at least in terms of size, since 35 percent of the national firms are also large; that is, they each produced more than 45,000 tons of paper in 1975. Size is related to rate of growth (Table 2), but only weakly so, and large firms are not the only ones to show strong growth trends. These findings may be compared with Katz's finding in the Argentine pharmaceutical industry that medium-sized firms grow the fastest.[26] On the other hand, Table 3 shows that national firms are found more often than transnational ones among the low-growth or no-growth categories: of seventeen national firms, 71 percent are in the last two categories, while two of three transnational firms are in the high-growth category. This finding is in agreement with Fajnzylber's conclusion that transnational firms in Mexico have a higher rate of growth than national ones.[27]

The degree of horizontal complexity, as measured by the number of internal divisions at the firm level, seems to be a characteristic of the industry, rather than one exclusively associated with the ownership of the firm (Table 4). Of twenty firms, eight have between four and seven divisions. The basic divisions to be found are generally finance, production, personnel, and sales. Table 5 shows that horizontal

TABLE 1

SIZE DISTRIBUTION AMONG TRANSNATIONAL AND OTHER FIRMS

| Ownership | Size of Firm* Large Firms (>45) | Medium-size Firms (10-45) | Small Firms (<10) | Total Firms |
|---|---|---|---|---|
| Transnational Firms | 2 | 0 | 1 | 3 |
| National Firms | 6 | 5 | 6 | 17 |
| Total Firms | 8 | 5 | 7 | 20 |

*Size grouped according to thousands of metric tons produced in 1975. Contingency coefficient = .26 (nonsignificant)

TABLE 2

RELATION BETWEEN SIZE AND RATE OF GROWTH

| Rate of Growth | Size of Firm* Large Firms (>45) | Medium-size Firms (10-45) | Small Firms (<10) | Total Firms |
|---|---|---|---|---|
| High growth (>6%) | 4 | 3 | 0 | 7 |
| Low growth (0-5.9%) | 3 | 1 | 2 | 6 |
| No growth or negative growth | 1 | 1 | 5 | 7 |
| Total firms | 8 | 5 | 7 | 20 |

*Size grouped according to thousands of metric tons produced in 1975. Contingency coefficient = .26 (nonsignificant)

complexity is related to size of firm, a correlation which agrees with reports on that relationship generally found in the literature. In that respect, transnational firms do not appear to behave differently from others; they tend to have large numbers of internal divisions, but no more so than their size category would suggest. When internal complexity is measured in terms of the number of professional specialties (Table 6), however, transnational firms stand out. While most firms tend to have mostly chemical engineers, accountants, and mechanical engineers, transnational firms usually include more

TABLE 3

OWNERSHIP AND RATE OF GROWTH

| Rate of Growth | Ownership: Transnational Firms | Ownership: National Firms | Total Firms |
|---|---|---|---|
| High growth ( >6%) | 2 | 5 | 7 |
| Low growth (0-5.9%) | 1 | 5 | 6 |
| No growth or negative growth | 0 | 7 | 7 |
| Total firms | 3 | 17 | 20 |

TABLE 4

HORIZONTAL COMPLEXITY OF THE FIRM

| Number of Divisions | Transnational Firms | National Firms | Total Firms |
|---|---|---|---|
| 1 to 3 | 0 | 6 | 6 |
| 4 to 7 | 0 | 8 | 8 |
| 8 or more | 3 | 3 | 6 |
| Total firms | 3 | 17 | 20 |

Contingency coefficient = .35 (nonsignificant)

TABLE 5

RELATIONSHIP BETWEEN HORIZONTAL COMPLEXITY AND SIZE

| Number of Internal Divisions | Size of Firm: Large Firms ( >45) | Size of Firm: Medium-size Firms (10-45) | Size of Firm: Small Firms ( <10) | Total Firms |
|---|---|---|---|---|
| 8 or more | 3 | 5 | 0 | 8 |
| 4 to 7 | 0 | 3 | 2 | 5 |
| 1 to 3 | 0 | 3 | 4 | 7 |
| Total firms | 3 | 11 | 6 | 20 |

Contingency coefficient = .55 ($p < .05$)

engineering specialists and more business school graduates.[28]

Transnational firms also tend to excel in the academic level of their executive personnel and in the variety of their professions (Table 7). The proportion of personnel with university degrees on the first three levels of the managerial hierarchy is close to 100 percent in their case. By contrast, national firms show a lower average.

When we asked the heads of firms and the heads of departments about the degree to which first-level executives participated in general policy decisions, and whether heads of firms interfered in the daily functioning of departments, we found that Mexican managers had already acquired the dominant "democratic" managerial ideology of the United States (that had its historical roots in the school of human relations). Few Mexican managers like to admit that they do not let their collaborators participate in decisions. Tables 8 and 9 reflect this situation, although we see some variance when it comes to general policy decisions. In that case, half of the national firms are characterized as being centralized in their decision making.

The third attitudinal question, whether the head of the firm "exercised strong leadership," seems to have been less ideologically loaded and provided better discrimination. Nevertheless, it is doubtful whether the very vague concept of "leadership" may be construed as an indication that the head concentrates decision making in his own hands. On all three questions, we find that transnational firms are less centralized. This impression is confirmed in Table 9, which gives the frequency of weekly formal meetings between heads of firms and first-level executives. In this case also, first-level executives in transnational firms have few formal meetings with the heads of firms. Since these meetings are programmatic rather than policy discussion sessions, fewer meetings indicate greater independence on the part of first-level executives.

Transnational firms also appear to be more formalized than other firms with respect to three items used as indicators of that characteristic (Table 10). Formalization should not be confused with the notion of "bureaucracy" as pathological. In fact, we found more indications of bureaucratic rigidities in firms that were not formalized in terms of the indicators defined. In one of these firms, for example, no item, no matter how minor, could be obtained by any employee without a formal request signed by the head of the firm. Generally speaking, firms that have gone to the trouble of writing a manual of procedures or systematic personnel evaluation are precisely those that have tried to simplify such procedures and implement a more rational system of internal communications. The higher degree of formalization of transnational firms is, in fact, an

TABLE 6

NUMBER OF PROFESSIONAL SPECIALTIES WITHIN THE FIRM

| Number of Professional Specialties | Transnational Firms | National Firms | Total Firms |
|---|---|---|---|
| 0 to 5 | 0 | 11 | 11 |
| 6 to 10 | 3 | 4 | 7 |
| 11 or more | 0 | 2 | 2 |
| Total firms | 3 | 17 | 20 |

Contingency coefficient = .57 ($p < .03$)

TABLE 7

PERSONNEL WITH PROFESSIONAL DEGREES IN MANAGEMENT RANKS

| Percentage of Management Personnel with Professional Degrees | Transnational Firms | National Firms | Total Firms |
|---|---|---|---|
| 75% to 100% | 3 | 5 | 8 |
| 50% to 74% | 0 | 5 | 5 |
| 1% to 49% | 0 | 7 | 7 |
| Total firms | 3 | 17 | 20 |

Contingency coefficient = .46 (nonsignificant)

TABLE 8

DEGREE OF CENTRALIZATION IN DECISION-MAKING

| Degree of Centralization | Transnational Firms | National Firms |
|---|---|---|
| Head interferes on routine matters | 0 | 5 (29%)* |
| Head alone decides on general policy | 0 | 7 (41%) |
| Head exercises strong leadership | 1 | 14 (82%) |

*Percentage of firms responding positively

TABLE 9

FREQUENCY OF WEEKLY FORMAL MEETINGS BETWEEN HEADS OF FIRMS AND FIRST-LEVEL EXECUTIVES

| Number of Weekly Formal Meetings | Transnational Firms | National Firms | Total Firms |
|---|---|---|---|
| No meeting | 0 | 1 | 1 |
| One meeting | 3 | 7 | 10 |
| Two meetings | 0 | 9 | 9 |
| Total firms | 3 | 17 | 20 |

Contingency coefficient = .39 (nonsignificant)

TABLE 10

DEGREE OF FORMALIZATION OF PROCEDURES

| Degree of Formalization | Transnational Firms | National Firms |
|---|---|---|
| The firm has a manual of procedures | 3 (100%)* | 5 (29%) |
| Personnel is periodically evaluated | 3 (100%) | 9 (53%) |
| Personnel evaluation is done in writing | 2 (67%) | 6 (35%) |

*Percentage of firms responding positively

Contingency coefficient = .32 (nonsignificant)

expected correlate of their relative decentralization, since formalization has been recognized as a substitute for direct supervision.

Summarizing our conclusions on organizational configuration, we do find indications that transnational firms have distinctive characteristics in comparison with other firms in the same sector: they tend to grow faster and are more complex internally, more professionalized, less centralized, and more formalized. The first four characteristics permit us to hypothesize that they should come out higher on the innovative scale because, in the past, creative firms have been those that change more rapidly and are more decentralized.[29]

*The Technology of Production*

Sociological studies of organizations have developed a number of instruments to measure and compare the concept of technology.[30] The difficulty they have tried to resolve was to find a concept that might apply to all kinds of organizations and that might be equally successful in describing how a certain input is transformed into an output. The only school of thought that has resolutely opted for measures that apply almost exclusively to manufacturing firms is the English school, starting with Joan Woodward and continuing with the Aston Group, who developed the concept of operation technology.[31]

Woodward's division between small-batch, large-batch, and continuous process is too broad to be of use in a one-product industry such as the pulp and paper one. We find that 100 percent of the firms we investigated can be classified as having large-batch technology.

The Aston Group's conceptual scheme is more fruitful for the present purpose. In our study, the flow diagrams of each production line in all twenty-seven factories were analyzed in the following way: the ratio of automated operations to the total number of operations was calculated and standardized by the number of lines.[32] We thus obtained an index of automation varying in principle from 0 to 1, but in fact from 0.5 to 1.

A second concept central to the notion of operation technology is the nature of the links that unite the different operations in a workflow. A rigid workflow is one for which technology is nonadaptable and designed for a specific limited use working along an invariable sequence. This concept is particularly important in developing countries. Economists have been arguing that imported technology should be adapted to local conditions in the developing countries, but have failed to devise a universal instrument to measure the technological rigidities inherent or modifiable in the production process.

The economic parameters that confront manufacturing firms in the developing countries are notoriously different from those for which most technologies have been devised, with a resulting waste of resources and high production costs. In this study, we investigated some of the ways in which the technological characteristics of firms were adapted to domestic conditions, such as scarcity of capital, small capacity of the market, and abundance of manual labor.

As measures of adaptability to a narrow market, we investigated first, the capacity at which plants were functioning, and second, the degree of product diversification; these reflect the ability to compensate for the smallness of the market and maximize the use of machinery. As a measure of adaptability to capital scarcity, we noted the presence of used machinery and the age of machinery as measured by the number of years it had been in use. Finally, the number of metric tons produced yearly per worker served as a substitute measure of capital intensity.

Automation is usually high in this sector and has been for many years—as even very old equipment is automated. The basic processes are therefore uniformly automated. Variation occurs in the input to the production process (whether ingredients are put in manually or piped in with preprogrammed proportions in chemical components) and in the finishing of the product from cutting to sorting, weighing, packing, and conveying to the warehouse.

The transnational firms included in this study depart from the pattern that has been attributed to them: they are less likely to be fully automated than other kinds of firms (Table 11). And they are more likely to have rigid workflows than national firms (Table 12).

Utilized capacity in transnational firms is usually higher than in national firms, as Table 13 shows. Three of the four plants of transnational firms worked at high capacity, while fourteen of twenty-three national firms did so. Nevertheless, there is little association between these characteristics.

The number of product lines per plant may be thought to be mainly a function of size. Table 14 shows that, in effect, there is an association between size and number of products. By contrast, the distribution of the number of product lines according to type of ownership (Table 15) shows that transnational firms tend to be more diversified than national firms: 65 percent of the plants of national firms produce from one to four product lines, while all transnational firms produce five product lines or more. This result is significant in light of Jorge Katz's finding in the Argentine pharmaceutical industry that firms grow mostly as a result of launching new products in the market.[33] It also confirms the hypothesis that product diversification (when equipment is versatile) is a way of achieving full capacity operation in the presence of a narrow market.

The purchase of secondhand machinery seems to be a widely accepted custom in an industry where the lifetime of machinery (at least basic machinery) is very long. We found that transnational firms are somewhat less likely to buy secondhand machinery, unless they take over existing Mexican factories. In such cases, rather than buy new machinery, they modernize what is in the plant. The question of whether or not it is more economical to buy used machinery is not easy to decide. Among the smaller marginal companies that we visited, we got the impression that the age of the purchased secondhand machinery was so advanced that such firms could not possibly compete in cost with even

TABLE 11

DEGREE OF AUTOMATION BY PLANT

| Degree of Automation | Transnational Firms | National Firms | Total Plants |
|---|---|---|---|
| High (0.9-1)* | 2 | 17 | 19 |
| Low (0.5-0.8) | 2 | 6 | 8 |
| Total plants | 4 | 23 | 27 |

*Index devised by Aston Group

Contingency coefficiency = .18 (nonsignificant)

TABLE 12

RIGIDITY OF WORKFLOW BY PLANT

| Rigidity of Workflow | Transnational Firms | National Firms | Total Plants |
|---|---|---|---|
| High (21 or more)* | 4 | 12 | 16 |
| Low (1 to 20) | 0 | 11 | 11 |
| Total plants | 4 | 23 | 27 |

*Links on scale devised by Aston Group

Contingency coefficient = .32 (nonsignificant)

TABLE 13

UTILIZED CAPACITY BY PLANT

| Utilized Capacity | Transnational Firms | National Firms | Total Plants |
|---|---|---|---|
| 1% to 80% | 1 | 9 | 10 |
| 81% to 100% | 3 | 14 | 17 |
| Total plants | 4 | 23 | 27 |

Contingency coefficient = .11 (nonsignificant)

the medium-sized firms. Slowness of operation and frequency of breakdowns of machinery, as well as the generally low qualifications of workers in such firms, accounted for a great deal of lost time. One would need to devise a scale to serve as a guideline for determining whether or not to purchase secondhand machinery. Its specifications would necessarily relate to standards of performance in the industry as a whole.

Even though relatively few firms buy used machinery, they tend to keep equipment purchased new for a very long time. The exact mix of age of machinery is difficult to measure, since factories are usually patchworks of equipment coming from many countries and purchased at different times. Nevertheless, we were able to divide the equipment into four age groups and assign a proportion to the equipment of each age group owned by the respective types of firms (Table 16). Relatively few plants had a large proportion of equipment less than five years old. On the other hand, twenty of the twenty-seven plants had a large proportion of equipment over fifteen years old, some dating back to the nineteenth century. Transnational firms are somewhat less likely to have very old equipment (54 percent as against 75 percent for national firms), but are also somewhat less likely to have brand-new equipment. Nevertheless, these tendencies are not very marked, so no striking differences between transnational and other firms were ascertained.

Finally, the yearly production per worker (Table 17) shows a surprising distribution: transnational firms are found only in low and medium categories of that measure, while 40 percent of national firms have high production per worker. The explanation is that transnational firms tend to be more vertically integrated down to the finished product, while national firms are more apt to sell semifinished products. Since labor-intensiveness is found mostly in the finishing operations and does not add to production as expressed by weight, the more vertically integrated a plant is, the less production per worker. This finding indicates that national data concerning the creation of employment in the paper industry are distorted, since they do not include finishing operations done outside the industry, either by small marginal factories (*maquiladoras*) or by industrial customers, who do cutting and packing for their own purposes.

In conclusion, we may say that the technological characteristics of transnational firms in the Mexican paper industry do not parallel results obtained from previous studies in other manufacturing areas. They are neither more automated nor more capital-intensive; they are not apt to have higher equipment turnover than other firms. On the other hand, they have more product lines, and they work at higher capacity than national firms.

TABLE 14

NUMBER OF PRODUCT LINES PER PLANT BY SIZE OF FIRM

| Number of Product Lines Per Plant | Size of Firm | | | Total Plants |
|---|---|---|---|---|
| | Large | Medium | Small | |
| 1 to 4 | 7 | 3 | 5 | 15 |
| 5 to 8 | 2 | 3 | 1 | 6 |
| 9 or more | 5 | 0 | 1 | 6 |
| Total plants | 14 | 6 | 7 | 27 |

Contingency coefficient = .42 (nonsignificant)

TABLE 15

NUMBER OF PRODUCT LINES PER PLANT BY OWNERSHIP

| Number of Product Lines Per Plant | Transnational Firms | National Firms | Total Plants |
|---|---|---|---|
| 1 to 4 | 0 | 15 | 15 |
| 5 to 8 | 2 | 4 | 6 |
| 9 or more | 2 | 4 | 6 |
| Total plants | 4 | 23 | 27 |

Contingency coefficient = .42 ($p < .05$)

TABLE 16

AGE OF EQUIPMENT BY OWNERSHIP OF FIRM

| Age of Equipment | Transnational Firms | | National Firms | |
|---|---|---|---|---|
| | Number of Cases | Average Percentage of Equipment | Number of Cases | Average Percentage of Equipment |
| Less than 5 years | 3 | 22% | 7 | 47% |
| 5 to 9 years | 3 | 41% | 6 | 54% |
| 10 to 15 years | 2 | 14% | 6 | 40% |
| Over 15 years | 3 | 54% | 17 | 75% |

*Innovative Behavior*

Innovativeness is defined in two ways in this study. First, it means the capacity for invention of new procedures—that is, innovation in the strictest sense of the term. Second, it means the capacity to modify existing procedures and to introduce changes in the firm that may have been developed elsewhere but are new to the firm.

As previous conditions to innovativeness, we measured the degree of contact with technological information existing in the firm and the reputed attitude toward innovation of the board of directors. The first measure was an unweighted additive index of four items using the following questions: (1) Is the firm under contract with consultants or a research institution? (2) Does the firm have a licensing agreement with a foreign firm or does it receive technical assistance from abroad? (3) Does the firm have personnel trained abroad? (4) Does the firm have a contract with an equipment dealer? The attitude of the board of directors was measured by a direct question to heads of firms and heads of departments. Inventive capacity was measured by an unweighted index for three items composed of the following questions: (1) Does the firm act as consultant to other national or foreign firms (mostly in Latin America)? (2) Does the firm export technology? (3) Has the firm developed any process or product (even though it may not have been patented)?

A lesser level of innovativeness, but one that represents change in a firm, is the introduction of new products. In each case, we asked how many product lines had been introduced in the last five years and what concrete plans existed for new products. Lastly, we recorded changes made in equipment. These were grouped into four categories: (1) purchases of new equipment for new lines, (2) replacement of old by new equipment with better specifications, (3) replacement with modifications in the engineering details, and (4) new development.

Transnational firms were found to have more contacts with different sources of information, as Table 18 indicates. Fourteen of seventeen national firms received only one or two sources of information, whereas two of three transnational firms received three or four sources.

The index of inventive capacity in Table 19 shows that the overwhelming majority of paper and pulp manufacturing firms are not inventive: of twenty firms rated, fourteen showed no form of inventiveness at all. Yet, those few firms that engage in some form of inventive capacity are likely to be transnational firms. Among national firms, only 18 percent are rated above zero on the index, whereas two of three transnational firms are.

In the introduction of new product lines (Table 20), the same pattern

TABLE 17

PRODUCTION PER WORKER BY TYPE OF FIRM

| Production Per Worker* | Transnational Firms | National Firms | Total Firms |
|---|---|---|---|
| Low (30 to 50 tons) | 1 | 5 | 6 |
| Medium (51 to 100 tons) | 2 | 5 | 7 |
| High (101 to 200 tons) | 0 | 7 | 7 |
| Total firms | 3 | 17 | 20 |

* Measured in metric tons for 1975

TABLE 18

EXTENT OF TECHNOLOGICAL INFORMATION BY TYPE OF FIRM

| Type of Firm | Index of Technological Information | | | | Total Firms |
|---|---|---|---|---|---|
| | 1 | 2 | 3 | 4 | |
| Transnational firms | 0 | 1 | 2 | 0 | 3 |
| National firms | 8 | 6 | 2 | 1 | 17 |
| Total firms | 8 | 7 | 4 | 1 | 20 |

Contingency coefficient = .46 (nonsignificant)

TABLE 19

INVENTIVE CAPACITY BY TYPE OF FIRM

| Type of Firm | Index of Inventive Capacity | | | | Total Firms |
|---|---|---|---|---|---|
| | 0 | 1 | 2 | 3 | |
| Transnational firms | 1 | 0 | 2 | 0 | 3 |
| National firms | 14 | 2 | 0 | 1 | 17 |
| Total firms | 15 | 2 | 2 | 1 | 20 |

Contingency coefficient = .62 ($p < .01$)

TABLE 20

NEW PRODUCT LINES BY TYPE OF FIRM

| Type of Firm | Number of New Product Lines Introduced | | | Total Firms |
|---|---|---|---|---|
| | 0-1 | 2-3 | 4 or more | |
| Trans-national Firms | 0 | 2 | 1 | 3 |
| National Firms | 14 | 2 | 1 | 17 |
| Total firms | 14 | 4 | 2 | 20 |

Contingency coefficient = .54 ($p < .01$)

appears: fourteen of seventeen national firms had introduced either no new products or only one in the past five years, while all the transnational firms had launched at least two new products in that period.

Lastly, the extent of equipment change showed varying patterns (Table 21). Few companies have undergone no change at all. Plant enlargements are equally scarce (only ten of twenty-seven possible cases) and more likely to occur in transnational firms. On the other hand, the mere replacement of machinery without modification is characteristic of the majority of firms, but is also inversely related to size—it happens mainly in small and medium-sized firms. Original developments in processes or products are rare; when they do occur they are characteristic of large or transnational firms.

## Conclusion

Because of the limitations of the sample used in this study, the conclusion reached can only be indicative. The hypothesis set down at the beginning of this chapter was that the organizational characteristics of the subsidiaries of transnational corporations could account for the ways in which they differ from national firms in achieving innovation. The results obtained do not support that hypothesis with respect to organizational size, rather, other characteristics appear to be more significant. Transnational firms tend to grow faster, are more complex internally, and are more highly professionalized and less centralized than national firms of similar size.

TABLE 21

EXTENT OF EQUIPMENT CHANGE BY PLANT

| Type of Change | Transnational Firms | National Firms | Total Plants |
|---|---|---|---|
| No change | 0 | 3 | 3 |
| Changes for expansion | 3 | 7 | 10 |
| Machine replacement | 2 | 13 | 15 |
| Machine modification | 2 | 13 | 15 |
| Original development | 2 | 3 | 5 |

As for other organizational characteristics, the results are consistent with the hypothesis in that they account at least partly for the differences in innovative activities observed between national and transnational firms. The latter exhibit the organizational characteristics that have been recognized in organization theory as leading to greater innovativeness: complexity, decentralization, and change. Although these firms have less authoritarian structures (if such a term may be used), they have stricter formalized controls. In other words, their higher executives have superior methods of diagnosing continuously the potential sources of success or failure. The more traditionally oriented firms that rely on direct supervision and centralized decision making by the head of the firm have, in fact, less control. To direct an organization requires more than controlling people, and such ability is, in any case, limited.

A paradoxical finding is that the most technologically "dependent" firms (in terms of transfer of technology agreements with foreign firms) are, in fact, the more "innovative." This indicates that the goal of technological autonomy within the context of imported technology, as expressed by the official policy of the Consejo Nacional de Ciencia y Tecnología, is not entirely a utopian one.

We must admit, as most analysts have in the past, that Mexican firms in the sector studied are not very innovative as a whole. Yet it would be mistaken to assume that local sources of innovativeness do not exist. In this study we have found concrete examples of innovation other than the classical measures of expenditures on research and development. Many innovations in the form of adaptations to facilitate production and meet local circumstances were observed on the shop floor and were often unrecognized by top management as significant changes. There are probably other innovations of this type that will only be recorded by direct observation.

The sector under investigation turns out not to be an undifferentiated whole; innovativeness varies with the size of firm and the presence of transnational direction. It remains to be seen whether this pattern will be corroborated in studies of other industries.

## Notes

1. The term *transnational* or *multinational* has been defined in various ways in the literature. In particular, some authors insist that the term *transnational* should be used only in cases where such firms have a global strategy. In this study we make no such assumption. By a transnational or multinational firm, we simply mean a firm that has operations in more than one country.

2. Bernardo Sepúlveda and Antonio Chumacero, *La inversión extranjera en México* (Mexico, D. F.: Fondo de Cultura Económica, 1974).

3. Fernando Fajnzylber and Trinidad Martínez Tarragó, *Las empresas transnacionales* (Mexico, D. F.: CIDE, 1976).

4. Loretta G. Fairchild, "Empresas manufactureras conjuntas norteamericanas y nacionales en Monterrey: una comparación en logros," in *Dinámica de la empresa Mexicana: perspectivas políticas, económicas y sociales,* ed. Viviane B. de Márquez (Mexico, D. F.: El Colegio de México, 1977).

5. Jorge M. Katz, *Importación de tecnología, aprendizaje e industrialización dependiente* (Mexico, D. F.: Fondo de Cultura Económica, 1975).

6. Fernando Henrique Cardoso, "Associated-Dependent Development: Theoretical and Practical Implications," in *Authoritarian Brazil: Origins, Policies and Future,* ed. Alfred Stepan (New Haven: Yale University Press, 1973), pp. 142–76.

7. The term *subsidiary* refers to fully owned as well as partly owned subsidiaries, i.e., joint ventures.

8. Miguel Wionczek, Gerardo Bueno, and Jorge E. Navarrete, *La transferencia internacional de la tecnología: el caso de México* (Mexico, D. F.: Fondo de Cultura Económica, 1974).

9. Manfred Nitsch, "La trampa tecnológica y los paises en desarrollo," *Comercio Exterior* 21 (September 1971):816–23.

10. Sepúlveda and Chumacero, *La inversión extranjera en México.*

11. The Registro Nacional de Transferencia de Tecnología was created in 1973 to register, examine, and approve the transfer of technology contracts that Mexican firms have signed. No transfer of technology contract in Mexico has any legal value unless it is registered in that institution. Mauricio María y Campos, "La política mexicana sobre transferencia de tecnología: una evaluación preliminar," *Comercio Exterior* 24 (May 1974):463-77.

12. Sepúlveda and Chumacero, *La inversión extranjera en México.*

13. Fajnzylber and Tarragó, *Las empresas transnacionales.*

14. Nacional Financiera, *La política industrial en el desarrollo económico de México* (Santiago, Chile: Comisión Económica para América Latina, 1972).

15. Richard J. Barnett and Ronald E. Müller, *Global Reach: The Power of the Multinational Corporation* (New York: Simon and Schuster, 1974).

16. Internal demand has not yet been satisfied. Should there be surplus production in the future, it is doubtful whether Mexican industry could soon enter the export market, since the man-hours required to produce a ton of paper are about eight times those in Canada and Sweden, according to the *Memoria Estadística* published by the Cámara Nacional de las Industrias de la Celulosa y del Papel.

17. The reasons for this situation are complex and politically entangled. By law, forestry and pulp-making are not vertically integrated in Mexico. Private pulp-making companies must obtain yearly concessions on woodland at the cost of long and uncertain negotiations with state-owned forestry companies, or the *ejidos* (collective farms) to which the land belongs. Two immediate consequences are a lack of incentive for paper-making companies to go into pulp-making (and therefore the dependence on imports), and a lack of a consistent policy and practice of reforestation in Mexico.

18. For more details on the differential technical processes and the history of process development, see W. Paul Strassmann, *Technological Change and Economic Development* (Ithaca, New York: Cornell University Press, 1968), Chap. 7.

19. There is even a flourishing contraband trade in waste paper coming from the United States (reputedly better than used paper in Mexico because it has not been recycled). A highly organized national network of waste paper collection and transportation to paper factories provides marginal employment to thousands of Mexicans.

20. *Maquiladoras* are small assembly plants that service manufacturing companies by making certain items for them or taking over the

finishing of their products.

21. One suspects that some of these purchases were made for fiscal reasons or to comply with the new law of Mexicanization that demands a high percentage of Mexican capital in all new investments.

22. Fajnzylber and Tarragó, *Las empresas transnacionales.*

23. This corresponds to the definition adopted by Fajnzylber and Tarragó. No separate analyses are presented for state-owned firms because only two such firms were included in the sample.

24. Charles Perrow, "A Framework for the Comparative Analysis of Organizations," *American Sociological Review* 32, pt. 1 (April 1967):194-208.

25. Tom Burns and G. M. Stalker, *The Management of Innovation* (London: Tavistock Publications, 1961); Perrow, "Framework for Analysis of Organizations."

26. Jorge M. Katz, *Oligopolio, firmas nacionales y empresas multinacionales* (Mexico and Buenos Aires: Siglo XXI, 1974).

27. Fajnzylber and Tarragó, *Las empresas transnacionales.*

28. Accounting is considered a university degree in Mexico. Historically, it existed long before business degrees and has served as a substitute for them, which explains why firms with an older age-structure do not have business school graduates.

29. Victor Thompson, *Bureaucracy and Innovation* (University, Alabama: University of Alabama Press, 1969); Jerald Hage and Michael Aiken, "Program Change and Organizational Properties: A Comparative Analysis," *American Journal of Sociology* 72, no. 5 (March 1967):503-19; and Peter M. Blau and W. Richard Scott, *Formal Organizations* (San Francisco: Chandler, 1962).

30. Perrow, "Framework for Analysis of Organizations"; William A. Rushing, "Hardness of Material as Related to Division of Labor in Manufacturing Industries," *Administrative Science Quarterly* 13 (September 1968):229-45; David J. Hickson, D. S. Turk, and Diana McPheysey, "Operation Technology and Organizational Structure: An Empirical Reappraisal," *Administrative Science Quarterly* 14 (September 1969): 378-97.

31. Joan Woodward, *Management and Technology* (London: Her Majesty's Stationery Office, 1958); Hickson, Turk, and McPheysey "Operation Technology and Organizational Structure."

32. A production line or product line was defined as one composed of very similar final products, for which the only variation was one of weight, color, or size to which the product was cut.

33. Katz, *Oligopolio, firmas nacionales y empresas multinacionales.*

# 10
# Limited Search and the Technology Choices of Multinational Firms in Brazil

*Samuel A. Morley*
*Gordon W. Smith*

Even as multinational corporations increase their activities in less developed countries (LDCs), doubts persist about the willingness and ability of these organizations to adapt to environments quite different from those in which the bulk of their expertise has been accumulated. The area of technology adaptation to labor abundance has probably been of greatest concern and of greatest potential impact on the host country.

As we show in Chapter 11, multinationals in Brazilian manufacturing have indeed switched to more labor-intensive methods.[1] However, most changes involved "scaling down" for the Brazilian market and usually replicated smaller-scale plants operating in high-wage countries. Adaptations to low Brazilian wages were of generally minor importance. Other authors have found similarly slight adaptations to cheap labor in LDCs.[2]

The explanation of these minor adjustments is the critical element for policy. Production planners in the multinationals we studied rationalized their decisions in terms of a limited range of engineering-efficient technologies.[3] In most cases their views can be characterized by nonhomothetic production functions subject to economies of scale. The isoquants are sharply curved, and the expansion paths are concave to the capital axis. If this picture is correct, policies designed to raise the labor intensity of multinationals will succeed mainly at the cost of engineering inefficiencies or lost economies of scale.

But there is an alternative explanation. Simply put, competitive

The authors would like to thank Richard R. Nelson, R. Hal Mason, and the editor of the *Quarterly Journal of Economics*, for the valuable comments and suggestions. This chapter is reprinted with changes from "Limited Search and the Technology Choices of Multinational Firms in Brazil," *Quarterly Journal of Economics* 91 (1977):263–87. Reprinted by permission of John Wiley and Sons.

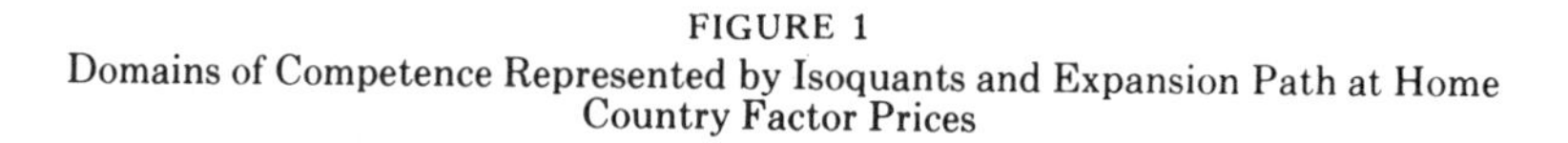
FIGURE 1
Domains of Competence Represented by Isoquants and Expansion Path at Home Country Factor Prices

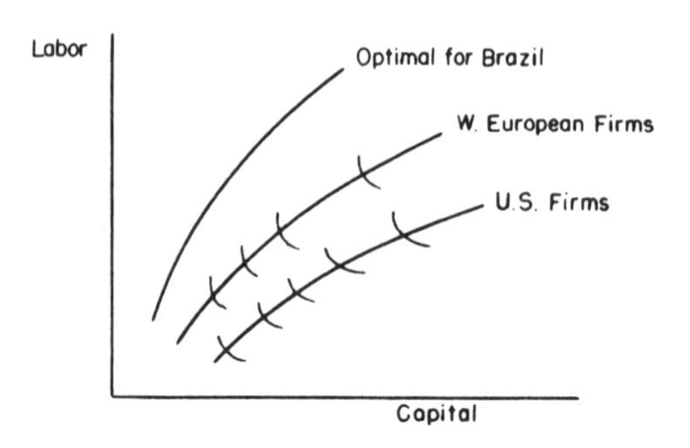


market pressures may be so weak that multinationals are able to meet their profit targets and other goals without searching for and employing the less familiar, more labor-intensive techniques. In this case, the full range of efficient capital-labor combinations is much broader than Figure 1 indicates, and measures that raise the competitive market pressures on multinationals (for example, import liberalization) may have a significant payoff in greater labor use.

This chapter investigates whether the failure of multinationals to adapt their production techniques more fully to labor abundance arises from truly limited factor substitutability or from a permissive environment, which allows multinational corporations (MNCs) to produce profitably without searching extensively for technological alternatives. We formulate a model in which search costs, accumulated experience, and a permissive environment play key roles in the choice of technique. The model is then used to analyze the differences in technique between firms of different nationalities operating in Brazil.[4]

If MNCs do not adapt more extensively to cheap labor mainly because of the technical features of production then we should expect to find little difference between the production methods used by firms of different nationalities. However, if lack of adaptation is due to limited search from initial positions of systematically different technological competence, significant and predictable variations between firms should be observed. Our tests, using four- and five-digit Brazilian establishment data for 1969, tend to support this second explanation.

The tests themselves do not permit a definitive interpretation of limited technological search. However, interview data suggest that "satisficing," or managerial discretionary behavior, probably plays an important role.

## Limited Search and Technology Transfer

The outlines of a model for the choice of technique are fairly simple. Assume a smooth production function for a given product. Furthermore, assume that, on the average, management differs across countries in its ability to combine labor and capital efficiently. Then we can write for a "typical plant" managed by an *i*th country firm:

$$Q_i = F(K,L,\lambda_i) \quad i = US, WE, BR = \text{country management}$$
$$= \lambda_i \, F(K,L)$$
$$\lambda_i = \text{the management efficiency parameter for the } i\text{th country.}$$

Casual empiricism of the Servan-Schreiber variety suggests that on the average, American firms may be more efficient operationally than West European firms, which in turn are almost certainly more efficient than the typical Brazilian firm. Thus, our working hypothesis is that

$$\lambda_{US} > \lambda_{WE} > \lambda_{BR},$$

where the subscripts have their obvious meaning.

At any time $\lambda_i$ may depend on the current capital-labor ratio. That is to say, management should be most efficient, given its accumulated experience, at existing capital-labor ratios, and the further $K/L$ departs from current levels, the lower $\lambda_i$ is likely to become. Similarly, $\lambda_i$ may depend on the average scale of operations. The further scale departs from that at which most experience has been accumulated, the lower $\lambda_i$ may become for that nationality. That is,

$$\lambda_i = f_i\left(\frac{K}{L}, Q\right)$$
$$\lambda_{i\text{-max}} = f_i\left[\left(\frac{\overline{K}}{L}\right)_i, \overline{Q}_i\right],$$

where the bars indicate current levels in the home country.

Now, the difference between $\lambda_i$ and $\lambda_{i\text{-max}}$ may be a declining function of the time elapsed since the introduction of the less familiar techniques. The long-run, static $\lambda_i$ may be independent of both scale and $K/L$. Sooner or later efficiency may reach the level previously achieved in the home country. It is perhaps more plausible that superior (say) American efficiency is partially inseparable from large-scale, mechanized operations. Fortunately, the issue is not crucial in this context.

The important point is that costs measured in terms of managerial inefficiency are greater, the greater the departure from tried and true methods. Furthermore, these costs are, by and large, unknown. The only way to determine them is through experience, although rough estimates could be made in principle.

To this should be added the search cost involved in discovering and mounting techniques differing substantially from those already employed by the firm. A rough approximation would assume that search and setup costs are an increasing function of the difference between the capital-labor ratio and scale to be "discovered" and those in which experience of the firm has been obtained. The payoffs to finding new labor-intensive techniques are of uncertain values and cannot be estimated accurately in advance.

If this discussion is correct, the set of factor combinations that can be employed by a firm with minimal search and managerial inefficiency, the firm's "domain of competence,"[5] is limited by its prior technological experience. The latter, in turn, should reflect an adjustment to past and present factor prices in the home country.[6] As a result, the "domain of competence" of American firms, which have long faced high relative wages, should be systemically more capital-intensive than that of European firms.[7] This hypothesis is pictured in Figure 1, together with the more labor-intensive *static* optimal technique for Brazilian factor prices.[8]

When a multinational considers setting up a Brazilian subsidiary, production planners will first examine techniques within or near the firm's domain of competence. For the permissive Brazilian environment, with its rapid growth, formidable import barriers, and lack of price competition,[9] this initial search will probably yield a fairly profitable technique. This will likely be more capital-intensive on average for American firms than for European.

Will search then continue? Three decision processes could lead to a cessation of search at this point:

1. Profit maximation under risk and uncertainty: the expected cost savings of new, more labor-intensive techniques are outweighed by the expected costs attached to uncertainty, search and setup, and managerial inefficiencies.[10]

2. The exercise of managerial discretion in a permissive environment: production planners are maximizing a function of variables other than profits.[11] Organizational slack, the "easy life" for management, *engineering* efficiency in production are all goals that would lead MNC production planners to less than optimal search from the stockholders' (and the host country's) viewpoint.[12]
3. Production planners may be *satisficing* in a permissive environment. Profit targets can be achieved with techniques from or near the firm's domain of competence, so production planners turn their attention to other problems.[13] Managerial discretion and satisficing are indistinguishable in leading to suboptimal search.

Whatever the explanation, search limited to the neighborhood of the domain of competence should lead to significant differences between technologies employed in Brazil by firms of different nationalities. For any given scale of output in Brazil $\overline{Q}$, we might expect U.S.-based firms to employ more capital per worker and obtain more value added per worker than the West Europeans. Note that the differences in efficiency parameters and capital-labor ratios affect relative profitability in opposite directions.

The situation of the Brazilian firm stands in strong contrast. Its domain of competence was developed in a low-wage, small market and most likely in *other product lines.* Therefore, although search costs may be higher for the Brazilian firm, if it is to be profitable, more extensive search is probably unavoidable. Furthermore, local firms probably believe they are operationally less efficient, and, therefore, they may give greater weight to the proper choice of factor proportions. For all these reasons, a Brazilian firm may be expected to employ a more labor-intensive technique, closer to the static optimum of Figure 1. This fact, combined with the lower managerial efficiency, would keep the value added per worker of local firms below the levels of multinationals.

We should note that the same predictions about capital intensities and value added could result if firms of different nationalities operated on the same production functions but faced systematically different factor prices. This identification problem is potentially serious only for comparisons of foreign and Brazilian firms. Small foreign subsidiaries are parts of generally much larger organizations with better access to capital markets than small Brazilian firms. In the larger-size categories, differences in capital costs should be less important. Indeed, large Brazilian firms have often enjoyed preferential access to official (subsidized) credit and have benefited from suppliers' credit for machinery imports. Duty-free importation of equipment has been the rule. We know of no serious study of the differential credit and exchange costs to domestic and foreign firms in Brazil. It seems reasonable to believe that even in the larger categories, the cost of funds to multinationals is somewhat lower than to Brazilian firms. Multinational firms may also pay higher wages than their local counterparts, but the quantitative evidence for this proposition is usually too aggregative for an accurate test. If the size of establishment, product mix, and quality composition of the labor force were to be controlled, it is not at all clear that significant differences in wages would be observed.[14] Indeed, of the seventy-four MNCs interviewed by Reuber et al., only thirty-two claimed to be paying blue-collar workers more in LDCs than the going rate, and only 17 percent paid more than 10 percent above market rates.[15] Thus, potential biases in the predictions introduced by systematic differences in wages seem slight.

Our direct evidence for limited search is of two types. First, from our interviews it was clear that multinationals did not engage in extended search for alternative techniques when planning their Brazilian operations. For example, in plant design, the thirty-five firms we interviewed in Brazil relied entirely on the experience in their American or European facilities. A broad search for new methods, particularly in low-wage countries, was never attempted.[16]

Second, some quantitative indication of the extent of search and the reliance of multinationals on home-country technology can be gleaned from figures on the sources of imported machinery by the nationality of the controlling firm. Data on projects over Cr$5 million approved in 1972 for duty-free importation of machinery and exemption from local taxes appear in Table 1.[17] The pattern is striking. Each nationality, except the Swiss and the French, imported more than 50 percent of its equipment from the home country. No nationality of foreign firms imported more than 28 percent of its equipment from other than home countries. The Japanese and Italians were particularly extreme in their concentration on their own equipment.

By far the most diversified importer was the Brazilian firm, probably an indication of greater search. (This fact is set forth above.) R. Hal Mason also found that locals procured from much more varied sources than U.S. firms in his sample of twenty-eight enterprises in the Philippines and Mexico.[18]

Although it is conceivable that this limited search may have resulted from profit maximization under risk and uncertainty, we suspect that satisficing managerial discretion also played an important role. Many of

TABLE 1

PERCENTAGE OF TOTAL MACHINERY IMPORTS TO BRAZIL ORIGINATING FROM VARIOUS COUNTRIES, BY NATIONALITY OF CONTROLLING GROUP: PROJECTS APPROVED BY CDI IN 1972 WITH TOTAL VALUE EXCEEDING CR$5 MILLION[a]

| Firm nationality | Country from which exported | | | | | | |
|---|---|---|---|---|---|---|---|
| | United States | Germany | Japan | France | Italy | Switzerland | Sweden |
| Brazil | 28.0 | 27.9 | 0.3 | 9.6 | 14.8 | 5.8 | 0 |
| United States | 60.3 | 22.6 | 0.4 | 0 | 3.6 | 0.2 | 0 |
| Germany | 10.2 | 70.2 | 0 | 0 | 2.3 | 12.5 | 0 |
| Japan | 7.8 | 7.0 | 78.2 | 0 | 0 | 3.4 | 0 |
| France | 27.4 | 27.5 | 0.9 | 35.1 | 5.6 | 3.5 | 0 |
| Italy | 10.7 | 1.7 | 0 | 0 | 82.4 | 4.8 | 0 |
| Switzerland | 0.6 | 95.6 | 0 | 0 | 0 | 3.8 | 0 |
| Sweden | 0 | 28.0 | 3.4 | 0 | 0 | 0 | 68.0 |

a. CDI, the Industrial Development Commission, approves duty-free import of machines.

*Source.* CDI project summaries as tabulated by IPEA/INPES, Industrial Sector.

the procedures revealed in our interviews are difficult to reconcile with profit maximization. For example, one aim of a diesel engine manufacturer was to make methods identical to those in [home country]. An auto maker claimed not to make detailed comparisons between the total costs of alternative types of machines in performing a task, but rather relied on experience in home country. A tractor maker claimed to be ignoring labor cost differentials entirely in choosing machines, since Brazilian wages were expected to be high as in [home country] in five years anyway.[19] One bearings manufacturer was aiming to use exactly the same machines in Brazil as in [home country]; another claimed to use "super modern" equipment, even more automatic than in [home country], a plus for the Brazilian subsidiary. A tire plant was said to be a virtual copy of the latest model in the home country. The list could be lengthened considerably.

Still, a definitive interpretation of limited search is impossible with the data at our disposal, but fortunately this is not crucial. The finding of differences in technique across firms would be enough to establish the highly significant fact that the range of efficient techniques is wider than each firm thinks it is.

## Empirical Results

The argument of the previous section suggests that the nationality of the firm may make a difference. Foreign firms at each scale should use more capital and be more efficient than Brazilian firms. To test this, we performed analyses of variance of value added per worker and a capital proxy, electrical energy per worker, in all Brazilian four-digit industries containing sufficient observations. Furthermore, American firms should be more capital-intensive and more efficient, on the average, than European firms of the same size. These and the previous hypotheses were tested at the Brazilian five-digit level, employing pairs of establishments as observations in sign tests. In all cases, we used the Brazilian government's Industrial Survey (the "Pesquisa Industrial") establishment data for 1969. Classification of manufacturing establishments by nationality of control was made according to *Guia Interinvest* (Rio de Janeiro, 1970). In general, any firm with at least 40 percent foreign control was considered foreign. Finally, only the seven states of Brazil's more developed Center-South were included.

### *Analyses of Variance: Differences Between Brazilian and Foreign Firms*

Formally, the analyses of variance were run using regressions with binary, i.e., "dummy," independent variables. In the value-added-per-worker regressions, three-way classification was used—size, capital intensity measured by electrical energy per worker (as explained below), and nationality; in industries where the number of observations permitted, interaction was allowed between that nationality and size effects. This permits the relative efficiency and capital intensities of foreign firms to vary with size. In the capital-per-worker regressions, the capital intensity cells are, of course, dropped from the list of independent variables.

$$(1) \quad \log \frac{\text{value added}}{\text{No. of production workers}} = \alpha + \sum_{i=2}^{n} \beta_i S_i + \sum_{j=2}^{m} n_j k_j + \lambda_2 N_2 + \sum_{i=2}^{n} \delta_i N_2 S_i$$

$$(2) \quad \log (k) = a + \sum_{i=2}^{n} B_i S_i + l_2 N_2 + \sum_{i=2}^{n} d_i N_2 S_i$$

with

$S_i$ = 1 if firm is in $i$th size category, 0 otherwise

$k_j$ = 1 if firm is in the $j$th capital-worker category, 0 otherwise

$k$ = capital-production worker

$N_2$ = 1 if the firm is foreign, 0 if Brazilian.

All dependent variables were transformed into logs. This assumes that the effect of each capital-intensity cell is a constant percentage across all size groups and nationalities, and that, when the number of observations does not permit interactions, the nationality effect is a constant in percentage terms across all capital intensities and size groups. Hence, the basic production function underlying the analysis of variance is assumed to be homothetic, except in the efficiency parameter when interaction terms are included. Particularly in light of our previous study, homotheticity does not seem warranted, since the capital intensity elasticity should depend on size.[20] In any case, no other restrictions were placed on the form of the underlying production function. The elasticity of substitution is free to vary with capital intensity, and returns to scale, if any, need not follow any prespecified functional form.

The coefficients of equation (1) are interpreted as follows: $\alpha$ is the log of value added per production worker of Brazilian firms in the smallest size group and the lowest capital intensity group. $\beta_i$ represents the increase in this figure due to membership in the $i$th ($\neq$ smallest size group), $n_j$ the increase belonging to the $j$th ($\neq$ lowest) capital intensity group, $\lambda_2$ the increase from being foreign, and $\delta_i$ the increase from being foreign and in the $i$th-size groups. A $t$-test of any given coefficient is equivalent to the $F$-test for the significance of the classification in classical analysis of variance.

The antilog of each coefficient is one plus the percentage that the average member of the specified cell exceeds the value added per worker of the small, low-capital intensity, Brazilian-controlled establishments. *All results are reported in antilog form* for ease of interpretation.[21]

The number of size and capital intensity cells we used naturally varied from industry to industry, depending on the range of these variables and the distribution of establishments across this range. Three size-groups were typical; five were the maximum. Three capital intensity categories were normally employed, although two and four occasionally were necessary.

Size of the establishment was measured by total value added in 1969. This is a better indicator of output than other variables that could have been used, such as the size of the labor force. But it must be admitted that systematic differences in product mix or monopoly power can disturb the relationship between "output" and value added. This is the more likely in the present context, since even the Brazilian four-digit level still contains considerable aggregation in some cases.[22]

Partially to alleviate this aggregation problem, and also to avoid including the technologically backward part of the (at least) dual economy,[23] all establishments with less than Cr$1 million value added in 1969 (about $250,000) were excluded entirely from the analysis. Most of the excluded observations were Brazilian. Presumably this exclusion also lessens considerably the identification problem introduced by differential finance costs.

Capital intensity, unfortunately, had to be measured in all cases by the value of electrical energy purchased per production worker. The 1969 Industrial Survey included no conceivable proxy for capital other than electrical energy. Suffice it to say here that electrical energy correlates quite well with installed horsepower and suffers the same basic shortcomings as a proxy of capital services.[24]

Finally, all governmentally controlled enterprises were excluded from our regressions.

The regressions were run for nineteen four-digit industry groupings, four of which (foundries and rolling mills) were grouped into two. Many important industries could not be included. Either the number of observations was far too small—auto making, tires, tobacco—or the sector was almost entirely Brazilian or foreign. Examples of the latter were, in addition to the previously mentioned three, textiles (Brazilian), food (Brazilian), and basic steel (public enterprises). As a result, the very largest foreign firms—Volkswagen, General Motors, Ford, British-American Tobacco, Mercedes-Benz, etc.—are absent from the regressions. (But they do appear in the sign tests reported below.)

In ten of the seventeen value-added regressions and nine of the electrical energy regressions, the $R^2$'s were significant at the 5 percent level or better (see Tables 2 and 3). In the remaining cases, size, foreignness, and our proxy for capital explain practically none of the variations across establishments. Apparently, there is a great deal of noise in the data, or other excluded variables, such as plant and equipment vintage, are important.[25]

In spite of this, there is support for the hypothesis that technologies differ. In eight of the nine significant "capital intensity" regressions, foreign firms used more electrical energy per worker than Brazilian firms when size was controlled, although in one case the foreign dummy was significant only at the 15 percent level. (See Table 4 for an illustrative cross-tabulation for auto parts.) Furthermore, the size-foreign interaction terms are significant at the 15 percent level in only one of the five industries in which they could be included, ferrous and nonferrous foundry products. There was usually no systematic difference in the foreign effect on capital intensity as size increased. This suggests that differential finance costs may not be an important factor influencing capital intensity, at least at the larger sizes included in our sample. It is also important to note that in none of the industries did foreign firms use significantly *less* electrical energy per worker than Brazilian firms. One suspects that the results would have been better with a more accurate measure of capital. In particular, foreign and large firms probably use more modern machinery on the avarage, so their capital services/electrical energy ratio is likely to be higher. Thus, our specification seems, if anything, biased against significance.

A similar picture emerges from the value added per worker regressions. In eight of the ten cases, the foreign efficiency parameter was significantly larger than one at the 10 percent level or better; in only one (vegetable fats and oils) was it significantly less than one. (See Table 4.) In electrical wires and cables, size and capital intensity were significant, but foreign firms were significantly more capital-intensive than their Brazilian counterparts.

The size dummies usually possessed considerable explanatory power, and in every industry at least one dummy was significant. In five of the

TABLE 2

SIGNIFICANT REGRESSIONS, VALUE ADDED PER PRODUCTION WORKER MULTIPLICATIVE FORM*
(Cr $ thousand, 1969)

| Variable | Industry | | |
|---|---|---|---|
| | Ferrous and nonferrous foundries | Nonelectric equipment for hydraulic, thermal, ventilation & refrigeration plants | Industrial machines and machine tools |
| Constant | 17.18 | 19.86 | 19.58 |
| Size group | | | |
| 2 | 1.305 (10%)[b] | 1.583 (5%) | excluded[a] |
| 3 | 1.204 (no) | 1.272 (16%) | 1.9333 (1%) |
| 4 | | | |
| Electrical energy | | | |
| 2 | excluded[a] | excluded[a] | 1.243 (2.5%) |
| 3 | excluded[a] | | 1.472 (2.5%) |
| 4 | excluded[a] | | |
| Foreign | 1.448 (5%) | 1.269 (10%) | 1.244 (6%) |
| Foreign-size interaction | | | |
| $F \times 2$ | 0.866 (no) | | 1.289 (10%) |
| $F \times 3$ | | 1.173 (no) | 0.702 (15%) |
| $F \times 4$ | | | |
| $R^2$ | 0.13 | 0.25 | 0.22 |
| *D.F.* | 54 | 50 | 115 |

| Variable | Industry | | |
|---|---|---|---|
| | Electrical wires and cables | Auto parts | Inorganic chemical products |
| Constant | 20.72 | 18.36 | 29.10 |
| Size group | | | |
| 2 | 1.677 (5%)[b] | excluded[a] | excluded[a] |
| 3 | | excluded[a] | 1.734 (2.5%)[c] |
| 4 | | 2.153 (0.5%) | |
| Electrical energy | | | |
| 2 | 2.425 (4%) | excluded[a] | excluded[a] |
| 3 | 4.548 (2.5%) | excluded[a] | excluded[a] |
| 4 | | | |
| Foreign | 1.049 (no) | 1.260 (4%) | 1.469 (2.5%) |
| Foreign-size interaction | | | |
| $F \times 2$ | | 1.278 (13%) | |
| $F \times 3$ | | 1.356 (3%) | |
| $F \times 4$ | | | |
| $R^2$ | 0.48 | 0.30 | 0.27 |
| *D.F.* | 13 | 123 | 35 |

TABLE 2 (*continued*)

| Variable | Industry: Vegetable fats and oils | Paints and varnishes | Pharmaceutical and medicinal products | Plastic prods. excluding those made from bakelite, ebonite, and galatite |
|---|---|---|---|---|
| Constant | 30.52 | 36.09 | 40.24 | 17.60 |
| Size group | | | | |
| 2 | 2.292 (0.5%)[b] | 1.227 (16%) | 1.538 (0.5%) | 2.361 (0.5%) |
| 3 | 12.83 (0.5%) | | 1.816 (0.5%) | 1.966 (0.5%) |
| 4 | | | 2.414 (0.5%) | |
| 5 | | | 1.545 (5%)[c] | |
| Electrical energy | | | | |
| 2 | excluded[a] | 1.504 (1%) | excluded[a] | excluded[a] |
| 3 | excluded[a] | | excluded[a] | 1.597 (1%) |
| 4 | | | | |
| Foreign | 0.5499 (5%) | 1.677 (0.5%) | 1.741 (2.5%) | 1.442 (2.5%) |
| Foreign-size interaction | | | | |
| $F \times 2$ | | | 0.674 (12%) | excluded[a] |
| $F \times 3$ | | | 0.869 (no) | excluded[a] |
| $F \times 4$ | | | 0.567 (6%) | |
| $R^2$ | 0.73 | 0.44 | 0.35 | 0.36 |
| *D.F.* | 25 | 29 | 111 | 65 |

*Predicted value for any firm = constant × product of coefficients for the classes in which it falls. "Excluded" = 1. E.g., predicted value for a foreign foundry, size group 3 = (17.18) (1.204) (1.448) = 29.95.

a. Not significant, = 1, excluded in final form estimated.

b. Number in parentheses is the (one-tail) significance level of the coefficient in the original log form.

c. All firms in this size group are foreign.

(When an entry is left blank, this means that the variable was not included in the regression.)

TABLE 3

SIGNIFICANT REGRESSIONS, ELECTRICAL ENERGY PURCHASED PER WORKER, MULTIPLICATIVE FORM
(Cr $1969)

| Variable | Industry | | |
|---|---|---|---|
| | Ferrous and nonferrous foundries | Industrial machines and machine tools | Parts and accessories for industrial machines |
| Constant | 174 | 224 | 175 |
| Size group | | | |
| 2 | 1.937 (2.5%) | 0.738 (2.5%) | 1.715 (2.5%) |
| 3 | 2.650 (5%)[b] | 1.196 (15%) | |
| 4 | | | |
| Foreign | 1.684 (7%) | 1.200 (7%) | 1.222 (15%) |
| Foreign-size interaction | | | |
| $F \times 2$ | 1.902 (15%) | excluded[a] | |
| $F \times 3$ | | excluded[a] | |
| $F \times 4$ | | | |
| $R^2$ | 0.30 | 0.07 | 0.23 |
| $D.F.$ | 54 | 117 | 29 |

TABLE 3 (*continued*)

| Variable | Industry | | |
|---|---|---|---|
| | Machinery & equipment for industrial and commercial installations | Electrical generators, motors and transformers | Pharmaceutical and medicinal products |
| Constant | 176 | 164 | 178 |
| Size group | | | |
| 2 | 0.861 (5%)[b] | 1.691 (10%) | 1.299 (7%) |
| 3 | | 1.773 (5%) | 1.619 (1%) |
| 4 | | | 1.430 (2%) |
| 5 | | | 1.727 (0.5%) |
| Foreign | 1.628 (0.5%) | 1.901 (1%) | excluded[a] |
| Foreign-size interaction | | | |
| $F \times 2$ | excluded | | excluded[a] |
| $F \times 3$ | | | excluded[a] |
| $F \times 4$ | | | excluded[a] |
| $R^2$ | 0.17 | 0.31 | 0.18 |
| *D.F.* | 43 | 37 | 115 |

| Variable | Industry | | |
| --- | --- | --- | --- |
| | Electric wire and cables | Television, radios, phonographs, tubes | Auto parts |
| Constant | 807 | 142 | 287 |
| Size group | | | |
| 2 | 0.787 (no) | excluded | 1.128 (no) |
| 3 | | | 1.643 (1%) |
| 4 | | | 2.268 (1%) |
| Foreign | 2.010 (2.5%)[b] | 2.045 (3%) | 1.271 (4%) |
| Foreign-size interaction | | | |
| $F \times 2$ | | | excluded[a] |
| $F \times 3$ | | | excluded[a] |
| $F \times 4$ | | | |
| $R^2$ | 0.29 | 0.18 | 0.23 |
| *D.F.* | 15 | 19 | 124 |

a. Not significant, = 1, excluded in final form.
b. Number in parentheses is the (one-tail) significance level of the coefficient in the original log form.
(When an entry is left blank, this means that the variable was not included in the regression.)

TABLE 4

CROSS TABULATION: AUTO PARTS

| Size group | Electrical energy per worker Cr $ | | Value added per worker th. Cr $ | |
|---|---|---|---|---|
| | Domestic | Foreign | Domestic | Foreign |
| 1 | 287 | 365 | 18.36 | 23.13 |
| 2 | 324 | 412 | 18.36 | 29.56 |
| 3 | 471 | 599 | 18.36 | 31.36 |
| 4 | 651 | 827 | 39.53 | 49.81 |

*Source.* Based on Tables 2 and 3.

ten industries, economies of scale were present even up to the largest size category. In four of these five—industrial machines and machine tools, auto parts, basic inorganic chemicals, and vegetable fats and oils—the coefficient of the largest size group is significantly greater than that of the immediately preceding cell.

Our capital intensity variable increased value added per worker significantly in only three of the ten industries—industrial machines and machine tools, electric wires and cables, and paints and varnishes. In these three, all energy coefficients were significant at the 4 percent level or better, and they increased in value with the "energy intensity" of the category. In plastics, one of two energy coefficients proved significant.

The most likely explanation for this state of affairs is again the crudeness of the capital measure, aggravated perhaps by a collinearity between size and foreignness and the "true" measure of capital intensity. That is, size and foreignness together may in some industries serve as a better proxy for capital intensity than electrical energy per worker does. Errors in measurement will bias downwards the electrical energy coefficients and lead to their insignificance.

The size effects, per se, are of minor importance for our purposes. But in general, the foreign parameters, which may "pick up" some of the capital effect, should be upper estimates of "pure efficiency" differentials, which probably explains their sometimes unreasonably large values.

The interaction terms were significant in only three industries—industrial machines and machine tools, auto parts, and pharmaceuticals. In auto parts, the value added of larger foreign establishments is relatively greater than that in the smallest size group, while in pharmaceuticals exactly the reverse occurs. There is no pattern at all in industrial machinery, and there is no conclusive evidence here as to whether foreign enterprise excels in larger-scale operations.

In summary, in about half of the industries examined, foreign firms used significantly more electrical energy per worker and attained a higher value added per worker when size was controlled. In the three industries where electrical energy per worker performs well as a capital intensity proxy, foreign firms appear to be more efficient than Brazilian firms on the average while in the other cases it is probably impossible to separate the efficiency effects of foreignness from its tendency to be associated with more capital-intensive processes.

### *Differences Between Foreign Firms*

Differential capital costs may still be cited as a partial explanation of the pattern we encountered in the previous section. Capital costs should be much less important in explaining variations across nationalities of foreign firms for two reasons. First, international enterprises of the type operating in Brazil face fairly similar capital costs. But even if they did not, our interviews suggest that multinationals are insensitive to capital costs in choosing their production techniques in Brazil. Hence, systematic differences between different nationalities of foreign firms may be taken as clearer evidence of satisficing managerial discretionary behavior.

In most four-digit industries, the number of foreign firms is too small for a separate analysis of variance. And the two-digit level is too aggregative for present purposes. Hence, we tried two other approaches. When data permitted, for each size category in each four-digit industry, we tested the hypothesis that the distributions of value added per worker and electrical energy per worker were stochastically at least as large for West European firms as for American. Rejecting this permits acceptance of the alternative that American firms show significantly higher values for both characteristics. The Mann-Whitney $U$-test was used. A summary of the results appears in Table 5.

Again, the evidence is mixed. Value added per worker differences are significant in the expected direction in only one-third of the size categories. The low figure reflects, in part, the small number of observations in most cells. In addition, this approach yields little support for the view that American firms use more electrical energy per worker than their European counterparts. If electrical energy were an accurate measure of capital intensity, this would be fairly strong evidence that American firms are more efficient in combining resources

TABLE 5

SUMMARY OF TESTS[a] FOR DIFFERENCES BETWEEN U. S. AND EUROPEAN FIRMS IN BRAZIL: 1969

| | No. of size categories in which U. S. was larger at 11.5%-level or better | No. of size categories in which W. Europe was larger at 11.5%-level or better | No. of size categories with no significant difference |
|---|---|---|---|
| Value added per production worker | 8 | 0 | 15 |
| Electrical energy per production worker | 3 | 0 | 20 |

a. Using Mann-Whitney $U$-test. Includes all size categories with at least two U. S. and two West European firms. The statistic $U$ is always an integer. Several size categories included four firms of one national group and three of the other. With these samples, the one-tailed probability of obtaining $U$'s of 1, 2, and 3 are 0.057, 0.114, and 0.200. 0.2 is clearly too liberal a significance level, while 0.057 seemed too stringent for our purposes. Hence we chose 11.5 percent.

than the Europeans in many of the industries examined. But energy varies so erratically in relation to value added that it is difficult to believe that large errors in measurement are not present.

Excessive aggregation may also create noise in the data. To deal more effectively with this and the problem of sample size, we tried our second approach. We paired establishments of two different nationalities in the same five-digit industry and of roughly the same total value added.[26] As many five-digit industries were included as the data permitted, but usually only the largest establishments could be sampled. Separate pairings were made between American and German, American and West European (excluding Germany), American and Brazilian, and West European and Brazilian firms. The sign test was used to determine whether the median value added per worker and electrical energy per worker differed significantly between nationalities in the direction predicted by the limited search hypothesis.[27]

A summary of the several pairings appears in Table 6. The picture here is more clear-cut, perhaps due to the greater disaggregation. In the value added per worker pairings, all of the differences run in the direction predicted and are significant (one-tail test) at the 5 percent level or better. With electrical energy per worker, the United States–West Germany pairings are significant (1 percent) in the direction predicted. But in the others, no significant differences were detected.

TABLE 6

RESULTS OF SIGN TESTS FOR SIGNIFICANT DIFFERENCES (ONE-TAIL TEST) BETWEEN NATIONALITIES AT THE FIVE-DIGIT LEVEL

| Number of pairings in which | Value added per production worker | Electrical energy per production worker |
|---|---|---|
| U. S. > Brazil | 23 | 14 |
| Brazil > U. S. | 4 | 10 |
| Ties | 2 | 4 |
| Significance | 1% | not significant |
| U. S. > W. Germany | 19 | 22 |
| W. Germany > U. S. | 10 | 5 |
| Ties | 0 | 2 |
| Significance | 5% | 1% |
| U. S. > W. Europe (excluding Germany) | 16 | 12 |
| W. Europe > U. S. (excluding Germany) | 6 | 9 |
| Ties | 2 | 3 |
| Significance | 5% | not significant |
| W. Europe > Brazil | 14 | 8 |
| Brazil > W. Europe | 5 | 11 |
| Ties | 0 | 0 |
| Significance | 5% | not significant |

Our tentative interpretation of the results is as follows. The significant differences in value added per worker probably represent both capital intensity and efficiency effects. The very spotty performance of electrical energy as a capital proxy is due primarily to its crudeness. Other studies, including a recent paper by R. Hal Mason, which contain comparative data on foreign and domestic firms in LDCs show greater capital intensity in the foreign subsidiaries.[28] We feel fairly confident that a similar result would have emerged more clearly for Brazil had better data been available.

Furthermore, if the significant differences in value added per worker were due only to the efficiency parameter, the profits of American firms should be higher than those of other nationalities, and the profits of Brazilian firms should be smallest of all. There is no evidence of this (see Table 7). If anything, published balance sheets for corporations in Brazil support the view that no significant differences in profit rates before taxes exist between the various nationalities in the activities we have examined. No great credence can be placed in these profit figures for a variety of reasons. Still, the figures are consistent with other evidence

TABLE 7

AVERAGE PROFIT RATES BEFORE TAXES OF CORPORATIONS,[a] 1973, BY NATIONALITY OF FIRM

| | Brazil | U. S. | W. Europe |
|---|---|---|---|
| Metals and machinery | 19.7 | 17.6 | 18.0 |
| | (32) | (10) | (9) |
| Autos and auto related | 18.8 | 22.2 | 15.9 |
| | (11) | (15) | (12) |
| Electrical equipment | 23.8 | 11.6[b] | 11.3 |
| | (7) | (6) | (11) |
| Chemicals, paper, and plastics | 14.0 | 8.6 | 7.6 |
| | (17) | (14) | (13) |

a. Only firms included in DEICOM's 1969 cadaster were included in the calculations. There are many fewer firms than establishments, of course.
b. Excludes Standard Electrica, S.A., with a loss of 162.6 percent of net worth. Numbers in parentheses are the number of firms included.
*Source of Original Figures. Brazil Report*, 1973 Edition (São Paulo, Visão, S.A. Editorial).

suggesting that value added per worker differences reflect a profit-increasing effect for foreign firms—greater efficiency—and a profit-decreasing effect—excessive capital intensity.

## Policy Implications

Definitive conclusions from the various strands cannot be drawn. However, the evidence suggests that the choice of technique is not as limited as it is often portrayed to be and that the failure of firms to adapt may well be the result of their limited search in a permissive environment rather than of technical factors. If this is correct, several policy implications follow.

(1) If LDCs want multinationals to employ more labor-intensive methods, they should be prepared to reduce the permissiveness of the environment. This means allowing greater competition from imports and avoiding "overkill" in granting favors to attract foreign firms.

(2) Once profits become more difficult, multinationals may be expected to pay greater attention to optimal production methods, to engage in greater technological search. They may also be more responsive to labor subsidies designed to bring market prices of factors more in line with their social opportunity costs.

(3) On the other hand, if LDCs are permissive and insist on developing industries that are capital-intensive by their nature—this seems to be the case with Brazil until now—they should not be surprised that multinationals do not employ more labor.

**Notes**

1. We interviewed the production planners or managers of thirty-five foreign firms in Brazil, mostly in various types of metal fabricating activities. In-depth comparisons were also made between eight U.S. plants and their Brazilian subsidiaries. For further details see Chapter 11.

2. For example, Wayne A. Yeoman found that most U.S. multinationals he examined (thirteen) transferred their production processes intact to LDCs. See "Selection of Production Processes for the Manufacturing Subsidiaries of U.S. Based Multinational Corporations" (D.B.A. thesis, Harvard Business School, 1968). Grant L. Reuber, et al. in *Private Foreign Investment in Development* (New York: Oxford University Press, 1973) report that of a sample of seventy-seven multinationals, fifty claimed not to have altered their production techniques *in any way* to conditions in LDCs.

3. Many economists have agreed. See, for example, Richard S. Eckaus, "The Factor Proportions Problem in Underdeveloped Areas," *American Economic Review* 45 (Sept. 1955):539–65; Christopher Clague, "Capital-Labor Substitution in Manufacturing in Underdeveloped Countries," *Econometrica* 37 (July 1969):528–37; Gerard K. Boon, *Economic Choice of Human and Physical Factors in Production* (Amsterdam: North-Holland, 1964). Gustav Ranis has strongly disagreed with this position in "Industrial Sector Labor Absorption," *Economic Development and Cultural Change* 21 (April 1973):387–408.

4. Brazil is excellent for this purpose. There is in its manufacturing sector a large foreign presence representing many nationalities. Furthermore, Brazilian labor costs per worker remain one-fifth their U.S. levels, according to the multinationals we interviewed in Brazil.

5. This term was suggested by Richard R. Nelson.

6. For truly multinational firms, this observation is less accurate. But even a number of these seem to duplicate home-country methods to a considerable extent in *all* their operations.

7. A most thorough study of international differences in factor proportions was performed by the Bureau of Labor Statistics. They calculated for 1964 the labor that would have been required by the British, French, and German iron and steel industries had they used American labor coefficients to produce their own output mixes. In each case, the figure was about *half* the labor the Europeans actually employed. Furthermore, the differences between the three European countries were exactly as one would have predicted, given their labor costs per hour. See "An International Comparison of Unit Labor Cost in the Iron and Steel Industries, 1964," *BLS Bulletin* 1580 (1968).

8. That is, abstracting from search and setup costs.

9. On these points see Joel Bergsman, *Brazil: Industrialization and Trade Policies* (London: Oxford University Press, 1970), Ch. 3; "Commercial Policy, Allocative Efficiency, and X-Efficiency," *Quarterly Journal of Economics* 88 (Aug. 1974):409–33.

10. A referee pointed out that fear of nationalization may increase risk and reduce expected return to such an extent that search is not extended. We found little indication of nationalization fears in Brazil, however.

11. See Oliver E. Williamson, *The Economics of Discretionary Behavior: The Managerial Objectives in a Theory of the Firm* (Chicago: Markham Publishing Co., 1967). "X-Inefficiency" is also relevant in this context. See Harvey Leibenstein, "Allocative Efficiency versus X-Efficiency," *American Economic Review* 56 (June 1966):392–415.

12. Louis T. Wells, Jr., in "Economic Man and Engineering Man: Choice of Technology in a Low Wage Country," *Public Policy* (Summer 1973):319–42, explains excessive capital intensity in a sample of Indonesian firms in part by monopolistic market power combined with the desire of engineers for mechanical efficiency.

13. Roy Radner, "A Behavioral Model of Cost Reduction," *Bell Journal of Economics*, 6 (Spring 1975):196–215, explores several models of satisficing behavior under uncertainty.

14. The Tariff Commission, for example, explained the wage differences it calculated between MNCs and Brazilian locals in manufacturing as a whole exactly in these terms. *Implications of Multinational Firms for World Trade and Investment and for U.S. Trade and Labor,* Committee on Finance, United States Senate, February 1973. However, R. Hal Mason in "Some Observations on the Choice of Technology by Multinational Firms in Developing Countries," *Review of Economics and Statistics,* 55 (Aug. 1973):349–55, did find significant wage differences in various skill categories between fourteen MNCs and fourteen Mexican and Philippine firms paired with them.

15. Reuber et al., *Private Foreign Investment in Development,* pp. 175–76. Curiously, only three of the thirty-two firms paying more than market scales were American.

16. Wickham Skinner found a similar lack of search of U.S. multinationals. See Wickham Skinner, *American Industry in Developing Economies: The Management of International Manufacturing* (New York: John Wiley and Sons, 1968), pp. 147–50.

17. It is fairly easy for a project to qualify—the set of industries

covered is very broad, and the criteria applied by the Comissão de Desenvolvimento Industrial are quite liberal.

18. R. Hal Mason, "Observations on the Choice of Technology by Multinational Firms," p. 352.

19. This implied a 40 percent per year rise in real labor costs in Brazil!

20. Computational problems in Brazil precluded the introduction of terms for the interaction of capital intensity and size. We were competing with the 1970 census for access to programmers and computer time in IBGE's data processing section. Only one pass at the regressions was possible, and nonhomotheticity in the capital variable was not part of it.

21. The significance levels, of course, are those of the original regressions.

22. There are 321 Brazilian four-digit industries compared with 422 U.S. four-digit industries.

23. See Richard R. Nelson, "A Diffusion Model of International Productivity Differences in Manufacturing Industry," *American Economic Review* 58 (Dec. 1968):1219–48.

24. Griliches used installed horsepower to estimate U.S. manufacturing production functions for 1954. Although its coefficients were highly significant, they seemed to be biased downwards because of measurement errors. See Zvi Griliches, "Production Functions in Manufacturing: Some Preliminary Results," in *The Theory and Empirical Analysis of Production*, ed. Murray Brown (New York: National Bureau of Economic Research, Columbia University Press, 1967), pp. 304–07. In the 1950 Brazilian Census of Manufactures (*Censo Industrial*), the rank correlation of electrical energy purchased per worker with installed horsepower per worker across ten size categories in the metal working industries was 0.96. Furthermore, in the U.S. Census of Manufactures, the rank correlation at the two-digit level between the rates of growth, 1954–1962, of electrical energy purchased and installed horsepower was 0.85. Neither measure includes buildings, but their principal defect seems to be their identification of *power* with capital *value*, clearly incorrect for sophisticated modern equipment. But book value as a measure of "capital" also suffers severe defects, all the worse in an inflationary environment.

25. Griliches and Ringstad, also working with disaggregated establishment data for Norway, obtained equally "pitiful" $R^2$'s. See Zvi Griliches and Vidar Ringstad, *Economies of Scale and the Form of the Production Function* (Amsterdam: North-Holland, 1971), Ch. 4.

26. Each observation was paired with another, the value added of which was most nearly equal to it. In only one case did a pair differ in

total value added by a factor greater than two. Usually the differences were much smaller.

27. See Sidney Siegel, *Nonparametric Statistics for the Behavioral Sciences* (New York: McGraw-Hill, 1976), pp. 68–75, for a good description of the test.

28. See, for example, Mason, "Observations on the Choice of Technology by Multinational Firms"; Loretta Louise Good, *United States Joint Ventures and National Manufacturing Firms in Monterrey, Mexico: Comparative Styles of Management* (Cornell University Latin American Studies Program Dissertation Series, no. 37, August 1972), p. 135; Wells, "Economic Man and Engineering Man," p. 323.

# 11
# Adaptation by Foreign Firms to Labor Abundance in Brazil

*Samuel A. Morley*
*Gordon W. Smith*

Of all the less developed countries, there is not one that has depended more on foreign direct investment for the development of its manufacturing sector than Brazil. As we pointed out in an earlier study, foreign investment played a key role in the import substitution phase of Brazilian industrialization before 1964, and this role has, if anything, expanded during the recent "Brazilian miracle."[1] Judged by increases in industrial output, the Brazilian growth strategy has been a phenomenal success. Manufacturing output has been increasing at a compound annual rate of 8.3 percent per year since 1949. Despite this impressive performance, critics have never been happy with the presence and the performance of foreign manufacturing firms in Brazil or in other developing countries.[2] They have argued that multinationals produce mainly for the upper middle class and contribute little to improving the lot of the poor. Multinationals are seen as an important contributor to the labor-saving bias of manufacturing growth, because of their importation of capital-intensive techniques of production, which are appropriate in their home countries, but are poorly adapted to the labor surplus economies of the typical developing country.

This is an issue that has generated a surplus of rhetoric but very few empirical studies. This chapter is an attempt to fill this void.

We seek answers to three basic questions: (1) To what extent have multinational firms adapted their production techniques to employ more labor and less capital in Brazil? (2) If adaptation has occurred, what are the principal reasons for it? and (3) Could the Brazilian government have induced foreign firms to seek more labor-intensive techniques than

This chapter is based on an article which appeared as "The Choice of Technology: Multinational firms in Brazil," *Economic Development and Cultural Change* 25 (January 1977):239–64. Reprinted by permission of The University of Chicago Press.

the ones they chose by manipulating effective factor prices or using other inducements for the use of more labor and less capital in production?

Brazil is perhaps the ideal country in which to try to answer these questions. Foreign participation in Brazilian manufacturing is large and well dispersed across industries and among nationalities.[3] Foreign firms have had sufficient time to overcome break-in problems, since the great majority of them set up operations before 1960. Brazilian labor costs are far lower than those in the developed world, something like one-fifth U.S. levels. Thus, the firms have had both the incentive and the time to adapt their techniques—if multinationals do adapt to conditions in less developed countries (LDCs), we should be able to observe it in Brazil.

Aside from numerous attempts to investigate factor choice econometrically, the most useful investigations for our purposes have been a number of recent case studies of the choice of technique by multinationals in their overseas operations.[4] The general consensus of most of these studies is that foreign firms make few adaptations of their production process when they move to LDCs. For Mexico and the Philippines, Mason found that foreign firms did attempt to conserve capital, but did not devote much effort to adapting existing technologies to local conditions.[5] However, this was due to the existence of small-scale equipment, which is apparently appropriate to the LDC market conditions. Yoeman found that most of the multinationals he interviewed transferred their production process intact.[6] This was particularly true in industries having oligopolistic market structures or where quality considerations limited machine choice. Grant Reuber and Louis Wells support this finding with two studies of multinationals in Indonesia.[7] Somewhat conflicting evidence is contained in a study by von Bertrab-Erdman.[8] He compared the operations of European firms in Europe and Latin America and found that adaptation was substantial, particularly in what he called the transport function, which is essentially material handling. These adaptations were not due to cheap labor in the LDCs, but rather to their smaller market size.[9]

In this chapter we seek to determine the extent of technological adaptation by multinationals in Brazil and to explain the factors primarily responsible for the types of adaptations that have occurred. Aside from data from secondary sources, we have generated our data and our impressions from interviews in Brazil of thirty-five foreign firms of different nationalities regarding the technological choices. Also we made in-depth visits to the manufacturing operations of nine metal working firms in the United States and their subsidiaries in Brazil. These form the basis for a more detailed description of technological adaptation in the metal fabricating industry.

Before we give evidence on the extent of adaptation by foreign firms to abundant labor in Brazil, we discuss the importance and the influence of prices and output volume on the choice of technique. Finally we shall discuss the policy implications of the evidence.

## Technological and Economic Determinants of the Choice of Optimal Technique

One of the difficult questions facing anyone who investigates the determinants of the choice of technique is to disentangle the influences of scale and factor costs. In the opinion of the businessmen we interviewed, size of market or scale was, by all odds, the most important determinant of their choice of production technique. Presumably, what that means is that at different scales of operation they see different techniques of production as dominating, almost regardless of factor prices.

In order to see why this might be so, let us consider the general problem of choice of technique at the micro-level. This choice follows in a straightforward way from cost minimization by the firm. In the standard analysis, the total costs of production for the various techniques in question are divided into fixed and variable elements. A functional relationship between output and variable cost is assumed. Hence, for any output level, the manager can calculate the total cost of production for each technique and choose the one with the lowest total costs.

The analytics of the problem for a simple choice involving only two techniques are illustrated diagramatically in Figure 1. $C_1C_1$ and $C_2C_2$ are the cost functions of the two techniques evaluated at market prices for the factors of production. $\overline{Q}_1$ is the capacity ceiling of technique one. It is the total output per year that can be produced using labor until its marginal product is zero.

Consider now how technique choice varies with changes in factor prices. What happens, for example, when we lower the wage rate and raise the cost of capital? Lowering the wage rate rotates each cost function in a clockwise direction, and raising the capital cost moves both functions upward by the same percentage amount. If we set the wage rate at zero and raise capital costs, we get the two new cost functions $C'_1$ and $C'_2$. Note that the change in factor prices only influences the firms' choice of technique in the range $Q_2\overline{Q}_i$. Technique choice is sensitive to reductions in the relative cost of labor if, and only if, the various possible techniques under consideration are all technically

FIG. 1.—Cost functions for alternative techniques

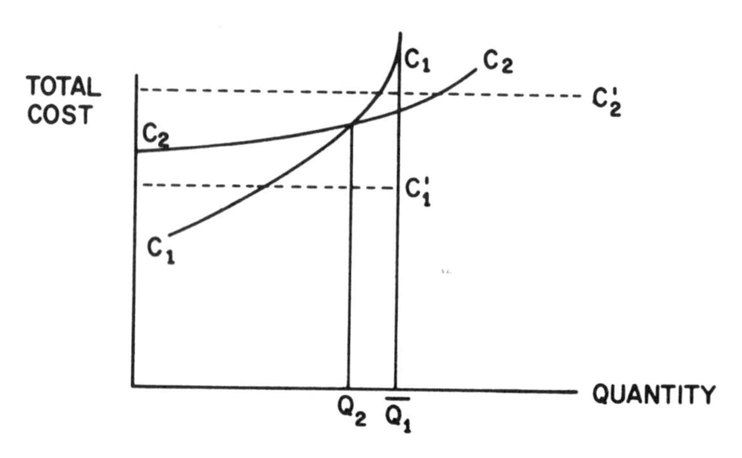


feasible at current output levels and if the more capital-intensive technique dominates at current relative factor prices. Diagramatically, the market price cost functions for the different techniques must intersect, and output must be between the intersection point and the capacity of the more labor-using technique. Price sensitivity of technique choice, therefore, depends on the size and position of what we will call the price-sensitive range (hereafter the PSR). Empirically, in many production processes, the PSR may turn out to be small or nonexistent. It may also turn out that it is at such low or high output levels that it is irrelevant. In either case, technique choice will be governed by scale. It will not be sensitive to factor prices.

Diagramatically, the PSR depends on differences in capital costs and the relative slopes of the cost functions of the different techniques. The greater the difference in capital costs and the smaller the difference in variable costs, the smaller the PSR range will be, or, to put it the other way, the greater will be the influence of output level on technique choice.

Let us, then, look more carefully at variable costs, the determinants of the slope of the cost functions. First of all, to the extent that the hourly output level of a machine or a technique is fixed and is invariant with respect to additions of labor, beyond the amount required to operate the machine, variable costs per unit of output will be constant and the cost function will be linear rather than convex. Other things equal, this will reduce the PSR, for it will mean that the businessman who wants to expand his output beyond the capacity of the smaller machine cannot do so by adding labor and speeding up the machine. His choice is either to buy a new machine with a higher capacity or duplicate his old process. If there are economies of scale in the machines themselves, the duplication option may be uneconomic at any set of factor prices.

Leaving the empirically important question of linearity versus convexity, what are the other factors influencing slope differences between different techniques or machines? Since the slope of the cost functions in Figure 1 is equal to labor and material costs per unit of output, the larger the differences in capacities of alternative machines—holding labor per machine fixed—or the greater the savings of operators in the more mechanized technique, the greater will be the differences between the slopes of the cost function, and the larger the probability of cost function intersection—a necessary condition for technique choice to be factor price-sensitive.

Turning now to capital costs, it is obvious that the greater the difference in cost between the more and the less mechanized technique, the smaller the probability of cost function intersection. Capital cost differences, in turn, are determined by lumpiness of capital. Are increments in capacity discontinuous and large? Where they are, the choice of capital is likely to be determined by the scale of operation and not by relative factor prices. This insensitivity will be increased if, in addition to large differences in capital cost between techniques, labor is a small component of cost even in the less mechanized technique. For this will make all the cost functions relatively flat and thus diminish the probability of intersection. For example, in the so-called process industries—chemicals, plastics, paper, cement, and glass—the main production process involves the transformation of some material, generally through a chemical process. Technical considerations require a substantial amount of machinery. In these industries, wages are never a very large cost of production except in ancillary services, such as shaping and packaging pills in the drug industry or moving logs into a pulp mill. Aside from the technical requirements of the main production process, one reason for this is that the material being handled is relatively easy to handle with machines. Either it is a liquid, in which case pipes and pumps dominate any more labor-intensive transportation system, or it is in regular shapes which are easy to design machines to handle. One might then predict that technique choice in the process industries would tend to be sensitive to the scale of operations and insensitive to factor prices.

A factor tending to increase the PSR is economies of scale in capital. For many different types of capital, it would appear that the cost of additions to capacity goes up less than proportionally. In the process industries, much of the capital is in the form of pipes, pumps, towers, and tanks used to hold and transport liquids and gases. It turns out that the costs of adding capacity in such vessels go up far more slowly than the increase in capacity because of the relation between interior volume and surface area of vessels.[10] Another form of economy of scale in capital involves the obvious but ignored fact that high capacity machines generally save space. Speaking loosely, where there are economies of scale in capital, cost function intercepts will be close together, so the price-sensitive range will tend to be large. This makes choice of technique sensitive to factor prices. That is because the more capital-intensive technique becomes the low-cost technique at relatively low output levels.

Yet even this conclusion needs to be qualified by further discussion. In the factories in the metal working industries that we visited, there were often many identical machines such as lathes, presses, or drills doing the same job. In such cases, the choice of technique meant whether to use a large number of low-capacity machines or a smaller number of high-capacity machines, or some combination of the two. If there are capital cost economies in such machinery, and if the capacity of each machine is technically more or less fixed, then variations in factor prices will affect only the choice of the marginal machine—not the entire machine stock. By marginal machines we mean the machines necessary to produce the output above even multiples of the most automatic machine. This says that if the output level is large enough to fully employ a set of one or more of the high capacity machines, then their use will be optimal regardless of the relative cost of capital and labor. For this conclusion to be necessarily true, labor per unit of output on the larger machine cannot be greater than on the small machine. Judging by our knowledge of labor requirements on various machines, that does not appear to be a very restrictive side condition.

The proposition that variations in factor prices affect only the choice of the marginal machine (for machinery fulfilling the conditions we specify) has the interesting implication that the larger the output of the factory, the smaller the elasticity of substitution between labor and capital. This is because the larger the factory, the smaller will be the percentage change in labor and capital implied by changes in marginal machines.[11]

The reason that the automatic machines dominate at the output level for which they were built is a result of the cost of increasing capacity and machine technology. When it is impossible to increase production on a machine by adding labor, it is impossible to match the production of the automatic by using more labor on the manual machine, regardless of how cheap that labor is. To produce an amount higher than the capacity of a manual machine, it is necessary to buy more manual machines and the space to put them in, and that is never profitable as long as the cost of the automatic rises at rates less than proportional with its increases in capacity. For particular sets of machines, therefore, an important empirical question will be relation between the output level of the typical factory and the capacity of the automatic machine. The bigger the first is relative to the second, the less sensitive labor use is going to be to relative costs.

To draw together this discussion of the various influences on the choice of technique, we have tried to differentiate between features of the production process that will increase the sensitivity of choice of technique to factor price variation and factors that will increase the influence of output level or scale and diminish the influence of factor prices. Features which increase the importance of factor prices are: (1) the ability to increase capacity limits by adding labor (this allows a firm to change its output level by adding labor rather than by buying more machines, which may be an attractive option where labor is cheap); (2) processes in which high-capacity machines have substantially lower labor requirements per unit of output; (3) economies of scale (the more similar the capital costs of various machines, the larger the PSR will be, particularly when the larger machine has a much higher output than the smaller. This conclusion must be modified where machine choice involves large numbers of identical machines. In these cases, where there are economies of scale, the choice probably is confined to the marginal machines, so the larger the factory, the less sensitive overall factor proportions will be to changes in factor prices.); (4) the more alternative machines or techniques there are, the wider the output range over which factor choice may be price-sensitive; this is a special case of (2) in the next paragraph.

Features which will tend to make scale of output the dominant factor determining technique choice are: (1) processes in which labor is an unimportant factor of production; (2) machines or processes in which increments to capacity are discontinuous and involve large increments in capital cost; (3) processes for which the output level of the factory is much larger than the output level of the most highly automated machine. With this discussion as a background, we turn now to the evidence on the extent of adaptation by U.S. firms in Brazil and the explanations for it.

## Adaptations by U.S. Multinationals to Local Conditions in Brazil

Do multinational firms adapt their production processes when they move to a LDC? Critics claim that they do not. We found little basis for such a claim. Indeed, the evidence suggests that adaptations by the multinationals to local conditions are substantial, at least in Brazil. In this section we present the evidence which leads us to make that statement. Our evidence is of three types: (1) data on comparative capital intensity from U.S. Department of Commerce surveys of U.S. multinational corporations; (2) a detailed comparison of metalworking machinery in the United States and Brazil, taken from machinery inventories in both countries; (3) direct evidence taken from plant visits in the United States and in Brazil of a sample of U.S. firms in the metalworking industries.

In Table 1 we show the most recent available data on the book value of fixed assets per employee in the subsectors of U.S. manufacturing and among U.S. multinationals in Brazil. While there are biases in these comparisons due to differences between book value and replacement cost and to differences in durability, there is little doubt in our minds that the large differences in capital per employee in the Brazilian operations of U.S. firms and in U.S. manufacturing primarily reflect real differences in production technique, rather than measurement error.[12] There may be some upward bias due to accounting differences, inflation, and durability, but this must be at least offset, if not more than offset, by our use of capital intensity averages for the industries in the United States. Since we know that the multinationals come from the largest size class within their subsectors, and since available evidence suggests that the largest size class uses more capital per employee than the average in practically every industry, true differences in capital per employee for U.S. firms in the United States and in Brazil may be even greater than Table 1 indicates. We also experimented with comparisons of value added per employee, capital services per employee, and capital services per dollar of output. Unfortunately, such comparisons are misleading where production functions are nonhomothetic or show economies of scale. If there are economies of scale, comparisons of value added per employee will mislabel as adaptation, differences in value added which are due to the higher output levels in the United States. If production functions are nonhomothetic because capital intensity increases with output, then value added per employee comparisons again will be biased in favor of acceptance of the adaptation hypothesis. Since we have every reason to suspect that both economies of scale and nonhomotheticity are present in the particular industries in which multinationals invest in Brazil, capital rather than value added per employee appears to be the best indicator of adaptation available.[13]

For purposes of completeness, we include in Table 2 the available data

TABLE 1

CAPITAL PER EMPLOYEE IN U.S. MANUFACTURING AND IN U.S. MULTINATIONALS IN BRAZIL

| | Employees, 1966 (Thousands) (1) | Net Fixed Capital, 1966 ($ Millions) (2) | Fixed Capital per Employee ($) (3) | Fixed Capital per Employee in U.S. Manufacturing, 1968 ($) (4) |
|---|---|---|---|---|
| Food production. | 15 | 44 | 2,820 | 13,165 |
| Paper........... | 3 | 20 | 6,452 | 25,872 |
| Chemicals....... | 26 | 235 | 8,333 | 35,202 |
| Primary and fabricated metals.. | 11 | 43 | 3,739 | 9,368 |
| Machinery....... | 13 | 59 | 4,338 | 10,488 |
| Electric machinery | 20 | { | 2,373 | 6,804 |
| Transportation equipment..... | 14 | 84 | ... | 8,993 |
| Other: | 25 | 86 | 3,307 | ... |
| Stone, clay, and glass........ | 8 | ... | ... | 19,320 |
| Rubber....... | 8 | ... | ... | 11,641 |
| Instruments.... | 5 | ... | ... | ... |
| Other......... | 4 | ... | ... | ... |
| Total manufacturing........... | 128 | 571 | 4,280 | ... |

SOURCES.—The distribution of employees by subsector of manufacturing in Brazil is taken from U.S. Senate, Committee on Finance, *Implications of Multinational Firms for World Trade and Labor*, 93d Cong., 1st sess., 1973, p. 707 (hereafter referred to as Tariff Commission Report). Net fixed capital by sector is reported in U.S. Department of Commerce, Bureau of Economic Analysis, *U.S. Direct Investments Abroad, 1966, Part II: Investment Position, Financial and Operating Data, Group 2. Preliminary Report on Foreign Affiliates of U.S. Manufacturing Industries* (Springfield, Va.: National Technical Information Service, 1972), p. 36. Fixed capital per employee in U.S. manufacturing is taken from U.S. Bureau of the Census, *Annual Survey of Manufactures, 1970 Industry Profiles*, M70 (AS)–10 (Washington, D.C.: Government Printing Office, 1972).

NOTE.—To obtain fixed capital per employee, we corrected the employment figures in col. 1 by the percentage difference between total employment reported in the Tariff Commission Report which was based on a sample of companies and the capital stock figures which are those of the complete universe of U.S. companies with operations in Brazil. From the 1966 census, p. 61, we have total employment of 133.5 thousand. To get total employees by sector we wrote up the figures in col. 1 by the percentage of employment reported in the census (133.5/128).

on value added per employee as reported in the U.S. Tariff Commission study of U.S. multinationals. The patterns appear to be roughly consistent with the fixed capital per employee data in Table 1. Both measures suggest significant differences between the U.S. and Brazilian operation of U.S. firms.

As another indicator of adaptation it is possible to make a comparison of the use of automatic versus nonautomatic machinery in certain subsectors of manufacturing in Brazil and the United States.[14] Clearly,

TABLE 2

VALUE ADDED PER EMPLOYEE IN BRAZILIAN AND U.S. ESTABLISHMENTS OF U.S. MULTINATIONALS, 1970
($)

| | United States (1) | Brazil (2) | Col. 2 ÷ Col. 1 (3) |
|---|---|---|---|
| Food products | 17,918 | 5,346 | 3.35 |
| Paper | 14,768 | 6,444 | 2.29 |
| Chemicals | 21,800 | 8,015 | 2.72 |
| Rubber | 17,748 | 11,803 | 1.50 |
| Primary and fabricated metals | 16,383 | 7,167 | 2.29 |
| Machinery | 15,505 | 8,720 | 1.78 |
| Electric machinery | 14,327 | 6,174 | 2.32 |
| Transportation equipment | 14,339 | 6,659 | 2.15 |
| Stone, clay, and glass | 15,378 | 6,491 | 2.37 |

SOURCES.—The U.S. Tariff Commission Report, table A-65, reports sales per employee for the U.S. and Brazilian establishment of the sample of multinational firms. Since not all of the sample have operations in Brazil, the two ratios are not completely comparable. The U.S. sales are converted to value added using the ratios in U.S. Bureau of the Census, *Annual Survey of Manufactures*. Brazilian sales are converted using the 1969 ratios from Departamento de Estatisticas Industriais, Comerciais e de Servicos, *Producão industrial 1969* (Fundacão IBGE, Rio de Janeiro, 1971).

the lesser use of automatic machinery would be a significant way in which adaptation to a cheap labor, small-volume economy would occur. In 1962, the Economic Commission for Latin America published an inventory of all the metal working machinery for the principle establishments in the metal fabricating industries in the state of São Paulo.[15] For each three-digit subsector of metals, machinery, and transportation equipment, this inventory gives a list of forty-four different metal cutting and forming machinery for the year 1960. A roughly comparable inventory is published for the United States every five years by the *American Machinist*. By fortunate accident, most of the sectors in Brazil for which the data are available are ones in which U.S. firms produce a significant portion of Brazilian output. Given the detail by machine, it is therefore possible to classify all the different machines as automatic or (for lack of a better word) manual and to compare the proportions of the two types in the United States and Brazil. This comparison is a fair proxy for the comparison that we would like to have between machine use in establishments of the same firms in Brazil and the United States, since most of the foreign firms in the industry were established prior to 1960. However, they were not all U.S. firms. We do not know the precise percentage of non-U.S. firms, but it is significant. Comparing the use of automatic machinery for Brazil and the United States may overstate the true differences in establishments of U.S. firms if automation in Europe is lower than it is in the United States. However, an offsetting source of bias stems from the fact that we are comparing the average for Brazil with the average for the United States. It is probably true that automation in the multinationals is greater than the average for U.S. industry.

Table 3 shows the data for the machinery comparison. The Brazilian data are for 1960, the U.S. data for 1958. We have grouped the subsectors for maximum comparability. The table shows what is clear to anyone who visits U.S. and Brazilian establishments of the same firm, namely, the far greater use of general purpose, nonautomatic machinery in the Brazilian metal fabricating industries. According to the table, Brazilian industry used two to three times as much manual machinery as did its U.S. counterpart. Where the U.S. firm uses gear hobbers, boring mills, turret and chucking lathes, and centerless grinders, each of which are labor-saving specialized machines, the Brazilian establishment tends to use universal lathes and milling machines and plain cylindrical grinders. Such machines are appropriate for the small product runs typical of the Brazilian metal fabricating industry because they are universal machines which can produce a wide variety of products with low set up costs. Many of the automatics can, too, but the set up costs are much higher on a per unit basis for small product runs.

All of the evidence that we have reported so far depends on more or less aggregated data, which may or may not be a good proxy for the data needed to test the hypothesis that firms did not adapt their production technology. In an attempt to overcome this shortcoming, we selected a small sample of U.S. companies with operations in Brazil from the metalworking industries. We then conducted extensive plant visits at both the establishment in the United States and the subsidiary in Brazil in order to compare in detail the differences in production in the two locations. We chose the metalworking industries because of their importance in industrial development and because we imagined, a priori, that here, rather than in the process industries, price-sensitive factor substitution should be possible. Since each pair of factories makes almost exactly the same products and is under the same management, this comparison is the most valid (and the most laborious) way to explore the extent of adaptation. In this section we will report briefly on the impressionistic results of the investigation.

In the metalworking industries, there are several distinct production activities that are useful to study separately. They are casting, stamping, metal cutting, plating and hardening, and assembly. The influence of scale and factor prices on the production technique in each of these activities varies widely because of their great differences in technical characteristics and requirements. For example, in certain processes,

TABLE 3

COMPARISON OF AUTOMATIC AND NONAUTOMATIC EQUIPMENT IN U.S. AND BRAZILIAN METALWORKING INDUSTRIES

| | Nonautomatics (Units) | Automatics (Units) | Total (Units) | Proportion of Nonautomatic Machinery (%) |
|---|---|---|---|---|
| Motor vehicles: | | | | |
| United States....... | 22,792 | 102,082 | 124,874 | 18.25 |
| Brazil.............. | 5,735 | 6,094 | 11,829 | 48.48 |
| Electric equipment: | | | | |
| United States....... | 34,374 | ... | 111,406 | 30.85 |
| Brazil.............. | 1,177 | ... | 1,871 | 62.91 |
| Household appliances: | | | | |
| United States....... | 5,799 | ... | 28,830 | 20.11 |
| Brazil.............. | 700 | ... | 1,212 | 63.52 |
| Community equipment: | | | | |
| United States....... | 15,293 | ... | 53,524 | 28.57 |
| Brazil.............. | 250 | ... | 365 | 68.49 |
| Special industrial and metalworking machinery: | | | | |
| United States....... | 79,797 | ... | 250,776 | 31.82 |
| Brazil.............. | 3,736 | ... | 4,787 | 78.04 |
| General industrial machinery: | | | | |
| United States....... | 34,848 | ... | 145,926 | 23.88 |
| Brazil.............. | 979 | ... | 1,627 | 60.02 |

mechanization may be so costly that it is not feasible even at U.S. output levels—that is, even U.S. output may be below the PSR. Here, we should expect to find minor differences between the U.S. and the Brazilian factory. This sort of failure to adapt has very different implications from the one where the U.S. firm installs expensive U.S. machinery even though a less capital-intensive methodology exists. When the critics talk about failures to adapt, they are thinking of the second and not the first reason for the factories being similar.

Given the influence of scale, it is useful to represent schematically the relationship of Brazilian and U.S. scale to what we called above the price-sensitive range. At output levels lower than the PSR, the manual, nonautomatic technique or machine dominates. Above it, the automatic will be chosen, despite variations in factor prices. Admittedly, this idea of a single PSR is an abstraction, since there may be a whole range of machines rather than just one. In that case, there will be several different PSRs. We will use the abstraction because it is a useful way of organizing our thinking about predictable differences between U.S. and Brazilian production techniques across industries.

We can conceive of five different relations between U.S. and Brazilian output and PSR (See Figure 2). Type A production processes are those in

FIG. 2.—Types A–E processes

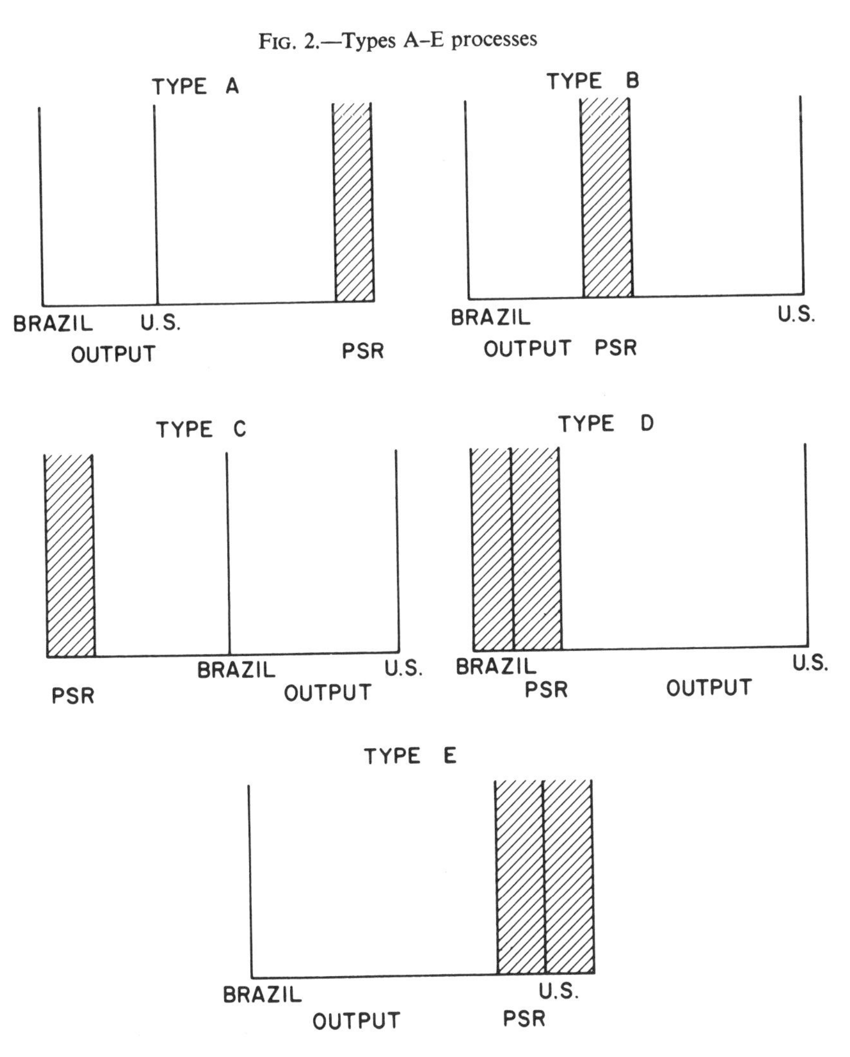


which the scale required before automatic equipment is profitable is so large that even establishments in the United States do not reach it. In Type B processes, the United States will be using automatics and the Brazilians will be using more labor-intensive equipment. It is in such processes that the greatest differences between machine and labor usage in Brazilian and U.S. establishments should be observed. In Type C processes, mechanization pays at relatively low levels of output, which are surpassed in both the Brazilian and the U.S. establishments. Type D and E processes are self-explanatory.

This representation undoubtedly overemphasizes the importance of output volume on the choice of technique, but it does so for the purpose of focusing attention on the tendency of adaptation of technique to occur only under rather restrictive technical and scale conditions. Only in Type B processes, and to a lesser extent D and E, should we expect to find significant differences in technique or adaptation by the foreign firm to local conditions. The critics of foreign investment who claim that the firms slavishly copy the techniques used in the United States without regard for differences in factor prices are implicitly assuming that production processes are all Type D, with the Brazilian establishments inappropriately using the mechanized U.S. technique. Whether this is true or not is an important empirical question.

We are now going to attempt to classify industries and their processes within metal fabricating in the scheme we have just developed. Processes were labelled A through C on the basis of our observations and classified as D or E on the basis of responses we got to two questions from the questionnaire: (1) how would production techniques in Brazil have differed had the relative cost of capital been 100 percent higher when the subsidiary was established? (2) how would techniques in Brazil be changed if output were at U.S. levels?

It is important to distinguish between two different types of industries in this sector, namely, those that produce at high and those that produce at low unit volumes. The former are standardized mass-production consumer durables like automobiles, radios, refrigerators, and the like, and the latter are mostly capital goods. Even in the United States, many capital goods firms produce so few units of output, and each unit is so highly individualized, that the firms could almost be characterized as machine shops producing to order.

Our research leads us to make the classification shown in Table 4 for the main production processes in metal fabrication for the mass-production, high unit volume type and the capital goods, low unit volume type industry.

We would now like to describe some of the specific establishment comparisons that led us to the conclusions contained in Table 4. The reader who is not interested in the details of lathes, acetelene torches, and transfer presses is advised to study the table and skip directly to the summary (p. 221).

Consider first the process of assembly. This is a heterogeneous but important group of activities involving the movement, handling, and assembly of parts. It is always labor-intensive, even in the United States; therefore, assembly lines in Brazilian and U.S. establishments are quite similar. The only exception to this observation is in small and relatively

TABLE 4

CLASSIFICATION OF PRODUCTION PROCESSES

| Process | Type | Differences between U.S. and Brazilian Production Methods |
|---|---|---|
| High unit volume, mass production industry: | | |
| Assembly | A | Small |
| Machining | B-C-E | Large |
| Pressworking | B | Large |
| Painting | B | Large |
| Welding and cutting | B | Large |
| Metal treatment | C | Small |
| Support services | E | Large |
| Low unit volume industry: | | |
| Assembly | A | Small |
| Machining | B | Large |
| Pressworking | A | Small |
| Painting | A | Small |
| Welding and cutting | B | Large |
| Metal treatment | B | Large |
| Support services | E | Large |
| High volume subassemblies: | | |
| Assembly | B | Large |
| Machining | B | Large |

simple subassemblies, which are produced in high volume typically for the automobile and electrical equipment industries. Mechanization, although technically feasible, is often very expensive and results in highly inflexible assembly lines. Any changes in model, part size, or design will generally require major modifications in the assembly machine. Thus, automatic assembly seldom pays, even at U.S. output levels and wage rates.

Even in the automobile industry, the high-volume industry par excellence, assembly is surprisingly labor-intensive. For a beginner, it is a great surprise to visit an automobile assembly line. Here, he imagines, is a scale at which he will find automation. And yet it is not so. He will not find machines putting wheels on axles or batteries in cars or drilling holes in chassis. A moment's reflection indicates why. A man has a really formidable comparative advantage over a machine in the assembly of something as complex and differentiated as an automobile because of his flexibility. Engineers do not have to invent a complex set of switches to tell him not to put truck tires on a Vega. Because of his flexibility, he is as good as an assembly machine and generally much cheaper per unit of output.

Where one does find automation is in small standardized subassemblies like brake drums, switches, transformers, and spark plugs.[16] Here, the number of parts to be assembled is small, fairly easy to handle, and most important, the subassembly is used on a wide range of different final products. Thus the subassemblies can take advantage of the economies of large scale, even though the final products suffer the diseconomies of product differentiation. For example, we saw an automated assembly line for disc brakes and brake parts at the Bendix Corporation in Michigan. On the disc brake line, six men assemble 450 brakes per hour using a sequence of special machinery which was designed and built for this particular disc brake alone. Nothing like that is economically feasible in Brazil at present output levels, since the Michigan plant makes as many disc brakes in one day as the Brazilian factory makes in one month. What this means is that the minimum scale of automated assembly techniques is larger than the scale of output of even U.S. firms throughout the metalworking industries, except in some components for the automobile industry and electric equipment. Thus, one should not expect great differences in assembly methods between establishments in the United States and Brazil.

Turning next to machining, we find a great diversity of machines, both in type and degree of automation. The basic operations are the cutting, grinding, drilling, and polishing of metal parts. For all the machines, a simple universal model is available on which the operator positions the work and selects and positions the cutting tool. Automation occurs either by some kind of programmed tool changes coordinated with position changes of both the tool and the work or by the use of transfer machines. These are, in essence, a series of universal machines, each programmed to perform a single task on a work piece which is automatically transferred from station to station. The small ones are circular, in the interest of saving space, while the large ones can be as long as 200 feet and have as many as 50–100 operations. In all of these machines, the automation is labor-replacing, although in some cases, the fact that a machine is automatically loaded allows higher cutting speeds or capacities as well.

A significant part of the fixed cost of all these automatic machines is their set up times. Any machine that does a sequence of operations, which can be varied for different parts, must be reprogrammed every time a new part run is started. This often takes a great deal of time and is only justifiable when the number of identical parts required is fairly large.

As Table 4 indicates, we found substantial differences between machining methods in U.S. and Brazilian establishments. Firms in the United States use far more automatics and special-purpose machines, particularly transfer lines, rotary indexing, and tape-controlled machines. Transfer lines are expensive, not only in terms of original cost but also in terms of maintenance. They require the services of highly skilled and costly electricians and toolmakers, as do all transfer equipment, such as transfer presses and tape-controlled metal-cutting machines. One thing most of the engineers in Brazil were emphatic about was the high cost of electricians and machine makers, relative to machine operators. Managers in Brazil were wary of using transfer machinery, even at U.S. market size. Where it is used, there were efforts made to adapt the machine to save skilled labor. For example, one firm managed to cut its tooling costs substantially by halving the number of stations on its transfer line and increasing the amount of metal cut per station. It did this by tripling the amount of time that the work piece stayed at each station. Output per hour fell, but there was a substitution of unskilled for skilled labor. These kinds of observations led us to classify these parts of machining as type E processes.

For many other machine tools, the cost of the more highly mechanized or specialized models does not appear to make them uneconomic even at Brazilian output levels. The scale at which a changeover to automatic equipment occurs is low, relative to high-volume industries, like automobiles or consumer durables. One finds automatic lathes, screw machines, automatic milling machines, and even small transfer lines in many of the Brazilian metalworking firms, even though their scale of operations is generally less than one-third of their U.S. counterparts. But these are firms in the production of consumer durables, where the volume of identical pieces of output is considerable.

In the capital equipment part of the sector, lot sizes are always very small, even in the United States. One sees many universal machines in U.S. factories. Automation takes the form not of indexing machines, but rather of individual machines which can be programmed to do a number of different cuts on a single work piece. The controls for the machine can be either electrical or mechanical—that is, tapes, computer programs, or systems of mechanical cams. These machines are obviously quite flexible, and the scale of output at which they become feasible is lower than that of transfer machines. At the U.S. establishments, all the capital goods producers whom we visited were using such machines, but there were none in Brazil. All the capital goods producers whom we visited in Brazil indicated that they would use about the same degree of automation as their U.S. parent at U.S. output levels, and we doubt that they would change this decision even if labor costs were substantially lower. Another factor which limits machine choice, particularly in the capital goods sector, is technical or quality constraints. Many capital goods require high precision machining or are of such large size that there may be only one machine that is technically able to make the parts.

This is why we classified machining in the capital goods sector as a type B process rather than a type E.

In metal stamping, the main type of automation is in loading and unloading devices for presses. These appear to require high output levels. In the United States, we observed floating die presses as well, but these do not yet exist in Brazil. The reason why automatic loading and unloading is not competitive with hand loading except at very high volumes is that it makes the press line rigid. Presses are, by themselves, universal machines, well suited to very small output runs. The operator has only to replace a set of dies. That obviously is not true with an automated set up, since the loading and unloading devices generally must be shape-specific. In other words, not only the dies but also the entire loading machinery has to be adjusted or changed for different parts. This increases the downtime and/or the fixed cost of the press. In Brazil, we did not find automatic loading or unloading or coil-fed presses, even in the automobile industry, which would be the first industry where volume might justify them. In the big consumer durables firms there seemed to be some question about press automation in the future. Plant engineers at both General Motors and General Electric felt that they would probably have to switch to automatics at U.S. scale because of space restraints, more than the direct cost advantage of the automatic over the manual press. Thus, both firms said that if labor costs were significantly lowered they would continue to use the manual presses. This is the reason we show pressworking as a type A as well as a type B process in Table 4. In the capital goods industry, output levels are so low that automation of presses does not appear to be widespread, even in the United States. Hence, we found little differences in the press shops of the U.S. and Brazilian establishments in this part of the metal fabricating industry.

Another general area that appears to be sensitive to both volume and factor prices is welding. Welding is an important input in any metal product made of sheet or stamping. Machines run the gamut from the hand-held acetelene torch to fully automatic electric-type welders. The question of significant automation in welding only arises in high-volume operations. At the present volume of operations in Brazil, we saw no automated welding set ups, although they are used in General Electric's refrigerator plant in Louisville and in the Chevrolet assembly plants we visited in Janesville. The reason is that welding machines are completely inflexible. To adjust a welding line for a new model is a far more complicated and costly proposition than to set up an automatic metal cutting machine. Therefore, unless model runs are very long, automatic welding cannot compete with hand welding. That is true even though the welding machinery itself is not terribly expensive. The expense is in the set up costs and in the downtime that would occur on small runs. As one might expect, there is relatively little difference between welding operations in Brazil and in capital goods firms whose output volume is small in the United States. As a matter of fact, Brazil appears to have a comparative advantage in products which are produced in small lots or quantities but which are made of welded rather than machined metal. Barber Greene, a maker of such a product—asphalt machines—is presently selling these machines all over Latin America and is considering producing all of its low-volume products in the Brazilian plant.

Painting is an area where high volume leads to significant changes in capital intensity. At U.S. volumes and U.S. labor costs, painting is quite highly automated in both the automobile and the consumer durables industries. Clearly, this is not true at output volumes in the capital goods industries. In Brazil, all painting that we saw was done by hand-held spray guns. There were no continuous baking furnaces for enamelled products. In several cases, firms had switched to electrostatic painting, but this was to save material, not labor. We found this lack of automation in Brazil, even in the automobile industry, somewhat surprising, since this is not a case where the automatic machine is highly inflexible, as it is in welding. The same set up would handle the painting of any model without costly changeovers. Apparently, low-cost labor is competitive with paint machines.

In what could be called services in support of basic production of the factory, we found the Brazilian plants using far more labor than their U.S. counterparts. This is because a great part of this activity involves material handling, an area in which labor has significant advantages over machinery. In Brazil, parts were generally moved by hand-pushed carts, and there was a far smaller use of conveyor systems either of the belt or the overhead type.

Another ancillary service at the factory is accounting, inventory control, and production scheduling. In the United States, a good deal of this is done with computers, while in Brazil, it is generally done by hand or with simple accounting machines. Even at U.S. volumes, this part of the operation will probably continue to be relatively labor-intensive in Brazil. However, production scheduling on assembly lines for products as complicated as cars probably cannot physically be done by hand, and the computer will prevail.

To summarize this section, all the evidence we have reported here points to a substantial modification of production processes by U.S. multinationals in Brazil. They use one-third to one-fourth as much capital per man, and the value added per man is less than one-half its U.S. level. Both personal observation and inventories of machinery confirm that Brazilian establishments of U.S. firms use fewer automatic and specialized machines than comparable establishments in the United States. They also tend to substitute labor for capital in what we have called the materials handling or support services of the production process, such as inspection, production scheduling, inventory control, and the like. This is not to say that they could not have gone further than they have in the direction of capital-saving, labor-using modifications. Evidence reported in Chapter 10 suggests that the firms do not search far for such modifications. Thus, it is possible that, disregarding search costs, adaptation by U.S. firms is not optimal. Nonetheless, we have no doubt that adaptation has been both substantial and significant.

### The Perception of Technological Alternatives: Evidence from Questionnaires

In order to determine why foreign firms had modified their production processes in Brazil, the authors personally conducted a survey in Brazil of thirty-five firms of several nationalities. The questionnaires took between one and two hours to complete, and answers were supplied by either the production methods planner or the vice-president in charge of production. When possible, plant visits were also arranged.

We made no attempt to attain a "representative" sample of foreign operations in Brazil. Firms in chemicals, pharmaceuticals, and paper were excluded as being either too secretive or not very interesting from a labor absorption view. Services and market activities were omitted entirely, even though their potential for labor absorption is great.

We tried to choose lines produced by firms of several different nationalities. Hence, the automakers, the television-radio assemblers, the telecommunications field, diesel engines, bearings, and tires were all included. In this way, it was hoped that significant differences between nationalities would be revealed. We also wanted to interview the largest foreign firms operating in metalworking activities, which meant again the automakers and also auto parts, and some machinery firms.

The technological portion of the interviews centered on the following questions:

1. Rank the following factors in terms of their importance in your firm's choice of production methods (processes, machines, etc.): quality of product; labor costs; shortage of skilled labor;

import licenses; cost and availability of financing; size of market; government incentives; others. Could you please give a specific example of how each of the three most important factors influenced the production methods chosen?

2. In the design of the plant, were the practices of similar firms elsewhere investigated? Where? In what respect?
3. Suppose machinery and equipment costs in Brazil (including imports) had been 100 percent higher in relation to labor costs/worker. How would this have changed the types and quantity of machines and the amount of labor employed, assuming no change in production levels?
4. Suppose production in Brazil were as large as an optimal plant in your home country. Would you then use the same types and quantity of machines as in your home country? The same quantity of labor? Explain.

With remarkable regularity, scale emerged as the overwhelming determinant of machine choice and labor use. Low labor costs in Brazil and/or the prospect of much higher machine costs were seen as having a small effect on factor proportions in the great majority of cases. Multinationals use more, sometimes three or four times more, labor per unit of output in Brazil than in their home country, but they say this is primarily a result of their smaller-scale operations in Brazil.

Table 5 presents the results of the rankings (question 1 above). Quality of product and market size are nearly tied for first place, while labor costs rank a distant third. Labor costs were ranked as often as they were only because several respondents believed that a reduction in labor use would be a prime way to *lower* costs.

Another indication of the importance attributed to scale is indicated by the answers to question 2. In many cases, the layout of a company subsidiary with a production volume equal to that expected in Brazil was a primary determinant in the design of the Brazilian operation. Yet we found no firms that had investigated production operations in low wage, as opposed to low output, level countries.

Confirming evidence on the importance of scale can also be seen in the answers to question 4. If current scale in Brazil were as large as in the home country, four of the interviewees said that machines and labor would be identical in *both* countries (see Table 6). Most often, however, firms saw certain very limited areas where differences would still persist because of cheaper labor. These sometimes involved sophisticated quality control equipment or superautomated machining operations which were worthwhile in the home country, but would not be in Brazil.

TABLE 5

RESULTS OF RANKINGS

| Factor | Rank 1 | Rank 2 | Rank 3 |
|---|---|---|---|
| Quality of product | 17 | 7 | 3 |
| Size of market | 14 | 14 | 2 |
| Labor costs | 1 | 8 | 9 |
| Shortage of skilled labor | 1 | 1 | 3 |
| Cost of finance | 0 | 2 | 3 |
| Government incentives | 0 | 0 | 3 |
| Import licenses | 0 | 0 | 1 |

NOTE.—Not all firms indicated all three first choices.

In many cases, the companies chose U.S. techniques because of the large physical volume of the plant that would be required to produce home-country volumes on universal-type machines. In essence, the firms were saying that they would substitute one type of capital, machinery, for another, buildings and land, both of which are quite expensive in São Paulo.

Only six firms interviewed claimed that significant differences would remain between the home operation and the Brazilian subsidiary at the same scale. Perhaps surprisingly, three of these were automobile companies. For example, Ford said that it would probably use the assembly methods currently employed in its European plants, and General Motors said that it would continue to hand-move material, hand-paint cars, and probably hand-weld.

Another way of approaching this issue is through question 3 above. As was to be expected, the answers closely paralleled those to the scale question (Table 5). Four of the thirty-one firms responding thought that a doubling of machine prices would bring no difference in machine and labor use. Most often, slight differences would be expected. But in a few cases, where automatic, higher volume equipment had recently been installed or was in the planning stage, higher machine costs would have made the moves uneconomical. For many companies, the magnitude of the adjustments was low because they were already using simple, universal machines in their metal machining, metal cutting, and press shops. Their techniques were already labor-intensive, and there was not much more that could have been done to substitute labor for capital. In terms of the analysis above, the majority of production processes appear to be types A through C, in none of which are factor prices of much importance in the determination of least-cost machine choice.

TABLE 6

HOW MACHINES AND LABOR USED IN BRAZIL WOULD DIFFER

| | From Home Country, If Scale Were as Large in Brazil | From Current Brazilian Subsidiary, If Machine Prices Were Doubled |
|---|---|---|
| No difference* | 6 | 4 |
| Slight differences* | 22 | 22 |
| Moderate to substantial differences* | 6 | 5 |
| No answer | 0 | 3 |
| Total | 34 | 34 |

* The answers were generally multidimensional except in the case of "no difference." The authors classified the other responses into these categories, based on their (partly subjective) views of their significance.

If this is correct, it means that the capital intensity of multinationals in Brazil was not a result of government policy incentives such as the importation of capital at subsidized exchange rates. The firms tended to replicate plants producing at the same scale elsewhere; it is doubtful that they would have modified plant design merely in response to a different set of factor prices. For the future, the main implication of these findings is that the expansions presently under way will lead most firms to alter their production methods in the direction of even greater capital intensity. Indeed, most of the firms we interviewed were in the process of rapid expansion to keep pace with the accelerated growth of the Brazilian market. In many cases, the facilities that are presently being built are, if anything, more modern than plants in the United States. What one sees at present in the Brazilian subsidiaries is the result of past decisions made for smaller market sizes. Current practice is labor-intensive by U.S. standards. Now that the Brazilian market has grown so large, actual production processes in use are often not least-cost and are being rapidly modified in the direction of increasing mechanization. Responses by the companies lead us to believe that, in most cases, those expansions and modifications would not have been altered to any great extent by changes in relative factor prices.

## Conclusions and Policy Implications

To summarize the findings of this study, we found substantial differences in production techniques between multinational firms and their subsidiaries in Brazil. However, these differences stem from scale differences, not cheap labor. At home-country output levels, most firms said that they would use home-country production techniques in Brazil despite the fact that the cost of labor is only one-fifth of what it is in the United States. Furthermore, most firms would not have modified their current operation very much, if at all, had the government made the relative price of labor much cheaper than it was.

When we looked closely at the way products are actually produced, we could see why production methods may be insensitive to relative factor prices. In the process industries, it seems clear that economies of scale and technical considerations dominate technique choice. Yet, even when we confine our attention to the metal fabricating industries, the same production features are present. In the first place, a good deal of the labor used in any factory is involved in either assembly or material handling. That part of production is labor-intensive even in the United States, and we found little differences between assembly methods in the two locations. Labor use in assembly will not respond to a reduction in the market wage rate because it is already intensively used in this function. In other parts of the production process, notably in machining and stamping, there are significant differences between the way things are produced in Brazil and the United States. But those differences seem to be explained more by the volume of operations in the two countries than by factor price differentials. For the high volume part of the industry, this is because the capacity of much of the automatic machinery is low relative to the annual production of factories, even in Brazil. Since the capital cost of machinery and space generally goes up less than the increases in capacity for this type of machine, it is likely that the use of high-capacity automatics will be cheaper than the alternative of using many low-capacity manual-type machines. In the capital goods sector, output is generally too small to justify the use of specialized machinery in Brazil. Production methods are now, and will continue to be, labor-intensive; we do not see much possibility for increases in labor usage, even if the relative price of capital is driven up. This conclusion is reinforced by quality constraints.

Having noted the arguments and evidence for low factor substitutability, we should point out that value added comparisons reported in Chapter 10 suggest that managerial perception of technical alternatives may also be an important element in technique choice, when the environment is permissive. Value added per worker and electrical energy per worker often differed significantly between foreign and Brazilian and U.S. and European firms. Furthermore, while multinationals did adapt to smaller scale, there is no evidence that they engaged in a serious search for more labor-intensive techniques than those employed in Europe or the United States.

A balanced conclusion, then, would be that scale dominates the considerations of the multinational firm and that the *behavioral* elasticity of substitution between labor and capital in most manufacturing is small. Thus, we are pessimistic about the amount of extra employment that would be created by multinationals in response to special measures, such as employment subsidies or increases in the cost of capital. Perhaps more direct insistence that multinationals seek more labor-intensive methods would have a significant payoff, but against this must be weighed the possibility that foreign investment might decline in response to government interference.

Is there, then, no solution to the employment problem, no way that industry can be induced to employ more workers? Our results do not imply that the situation is hopeless; they imply that people have been approaching the problem in the wrong way. The way any particular product is produced seems to be quite insensitive to variations in factor prices. Hence, once the pattern of output is determined, certain employment patterns are implied. To increase employment, one must start one step back—in the choice of products. Here, large improvements in labor absorption should be achievable by government inducements to produce labor-intensive products (rather than to produce any particular product in a labor-intensive manner).

Looking back at Brazil's industrialization drive, we would guess that the reason it produced so little employment was that the leading sectors were the mass-production consumer durables like automobiles and the process products like chemicals. These products are always produced in a capital-intensive fashion. To put it another way, poor employment results during the period probably should not be blamed on the fact that many of the leading firms were foreign, but rather on what those foreign firms were producing.

Labor-intensive products should be those produced at low volume and requiring much assembly or complicated materials handling, because these are the areas where human dexterity and flexibility are competitive with machinery. Along this line, it is interesting to note that several of the firms we interviewed in Brazil were planning to transfer many of their low-volume products from the United States to Brazil. Such developments should increase both exports and employment in industry at the same time. It seems to us that further progress in this area can be achieved by removing artificial inducements to the importation of labor-intensive products like machinery, by not giving special incentives for the establishment of mass-production products, and by continuing the export incentive program, since it tends to favor those labor-intensive products in which Brazil has a natural comparative advantage.

**Notes**

1. Samuel A. Morley and Gordon W. Smith, "Import Substitution and Foreign Investment in Brazil," *Oxford Economic Papers* 23, no. 1 (1971):120–35.

2. Theotonio dos Santos, "The Structure of Dependence," *American Economic Review* 60, no. 2 (1970):231–37; Osvaldo Sunkel, "The Pattern of Latin American Dependence," *Latin America in the International Economy,* ed. Victor L. Urquidi and Rosemary Thorp (New York: Wiley, 1973).

3. For an estimate of the foreign share in the subsectors of Brazilian manufacturing, see Morley and Smith, "Import Substitution," p. 128.

4. See M. Nerlove, "Recent Empirical Studies of the CES and Related Production Functions," *The Theory and Empirical Analysis of Production,* ed. Murray Brown (New York: National Bureau of Economic Research, 1967); and Zvi Griliches and V. Ringstad, *Economics of Scale and the Form of the Production Function* (Amsterdam: North-Holland Press, 1971).

5. R. Hal Mason, *The Transfer of Technology and the Factor Proportions Problem: The Philippines and Mexico,* United Nations Institute for Training and Research Report no. 10 (New York: United Nations, n.d.).

6. Wayne A. Yoeman, "Selection of Production Processes for the Manufacturing Subsidiaries of U.S.-Based Multinational Corporations" (DBA Thesis, Harvard Business School, 1968).

7. Grant L. Reuber, *Private Foreign Investment in Development* (Oxford: Oxford University Press, 1973); Louis T. Wells, Jr., "Economic Man and Engineering Man: Choice of Technology in a Low Wage Country," *Public Policy* 21, no. 3 (1973):319–42.

8. Hermann R. von Bertrab-Erdman, "The Transfer of Technology: A Case Study of European Private Enterprise Having Operations in Latin America with Special Emphasis on Mexico" (Ph.D. dissertation, University of Texas, 1968).

9. Ibid., p. 104.

10. For discussions and empirical estimates of economies of scale see David Granick, *Soviet Metal Fabricating and Economic Development: Practice vs. Policy* (Madison: University of Wisconsin Press, 1967); John Haldi and David Whitcomb, "Economies of Scale in Industrial Plants," *Journal of Political Economy* 75, no. 4, pt. 1 (1967):373–86; C. Pratten, *Economies of Scale in Manufacturing Industries,* Department of Applied Economics Occasional Papers, no. 28 (Cambridge, England: Cambridge University Press, 1971); C. F. Pratten and R. M.

Dean, *The Economies of Large Scale Production in British Industry,* Department of Applied Economics Occasional Papers, no. 3 (Cambridge, England: Cambridge University Press, 1965); and Aubrey Silberston, "Economies of Scale in Theory and Practice," *Economic Journal* 82, no. 1, suppl. (1972):369–91.

11. We came to this conclusion after studying particular examples of machine choice in metal and woodworking industries described by G. K. Boon in *Economic Choice in Human and Physical Factors in Production* (Amsterdam: North-Holland Press, 1964). He shows the different ranges of output and the different machine combinations that would be chosen at market prices for factors. He then sets wage rates at zero, triples the cost of capital, and compares the machinery used with that in the market price case. His conclusion is that factor substitution is greater at low output levels, which he thinks are those likely to be observed in LDCs.

12. On the problems of using book value of capital per employee as a measure of capital intensity, see Mason, *Transfer of Technology,* p. 25.

13. For evidence on the relationship between size and capital intensity, see Richard R. Nelson, "A 'Diffusion' Model of International Productivity Differences in Manufacturing Industry," *American Economic Review* 58, no. 5, pt. 1 (1968):1219–48; and Pratten and Dean, *Economics of Production in British Industry.*

14. For a previous use of this methodology, see Granick, *Soviet Metal Fabricating.*

15. Economic Commission for Latin America, *La fabricación de maquinarias y equipos industriales en América Latina: II Las marquinas herramientas en el Brasil* (New York: United Nations, 1962).

16. Von Bertrab-Erdman, "Transfer of Technology," pp. 91–95.

# 12
# Technological Fusion and Cultural Interdependence: The Argentine Case

*James H. Street*

The evolution of the Argentine economy during the past century represents an exceptional case of a relatively primitive society that was suddenly transformed into an advanced agricultural and commercial system and subsequently fell into a condition of persistent unsatisfactory growth. This condition has often been described as "economic stagnation," and, although there have been intervals of apparent improvement, the generally sluggish trend of the economy has been evident long enough to exhibit pronounced secular characteristics.[1] More recently, writers influenced by the Dependency School and the neo-Marxist concept of underdevelopment have tended to treat Argentina as a case of international economic dependency, stressing its critical reliance on foreign sources of investment funds, foreign exchange, and technology.[2]

A study of the country's history since its political independence indicates that the Argentine problem can best be understood in light of a theoretical framework derived from the work of C. E. Ayres and Simon Kuznets. Ayres, following Thorstein Veblen, clearly identified the cumulative process of technological innovation and its impact on social institutions as the prime source of economic progress through the course of human development. He drew particular attention to the evaluative judgments that distinguish technological from institutional behavior.[3]

While differing with Ayres regarding the causal interplay between technology and institutions, Kuznets, in his extensive empirical studies of modern development, nevertheless strongly reinforces the basic concept that a country's economic growth is "based on advancing

This chapter was published in an earlier form as an article under the title "The Ayres-Kuznets Framework and Argentine Dependency" in *The Journal of Economic Issues* 8 (December 1974):707–28. Permission to reprint is gratefully acknowledged.

technology and the institutional and ideological adjustments that it demands."[4] His major contribution in this context has been to generalize and to give empirical support to the structural changes that typically ensue when modern economic development begins.[5]

Viewed in the Ayres-Kuznets framework, Argentine development—notwithstanding occasional exogenous shocks—has been powerfully shaped by technological forces and institutional resistances. When Ayres speaks of "institutions" and Kuznets of "institutions and ideology," it is clear that both are concerned not only with formal structures, such as the policy making agencies of government, the educational system, and business firms, but also with conditioned attitudes and values that informally characterize the work ethic, inventive activity, and managerial practice, elusive as they may be to isolate. Institutional behavior, although altered by new experience, is unavoidably conditioned by past cultural influence.

The larger study on which this chapter is based raises two questions: Why did the Argentine economy grow so rapidly during the period from 1870 to 1914? Why did it thereafter show a chronic and increasingly widespread tendency toward long-term stagnation and technological dependency? The answers to the second question are strongly conditioned by the answers to the first.

## Evidence of Stagnation and Dependency

Economic historians generally agree that the period of most vigorous economic growth in Argentina occurred from about 1870 to the beginning of World War I in 1914. Statistics based on contemporary data are lacking, but Carlos F. Díaz Alejandro has estimated that the real gross domestic product grew at an average annual rate of at least 5 percent during the half century preceding World War I, which he describes as "one of the highest growth rates in the world for such a prolonged period of time."[6]

Growth accelerated during the latter part of this period. The United Nations Economic Commission for Latin America (ECLA) has estimated that during the period 1900–1914, the Argentine gross domestic product expanded at an annual rate of 6.3 percent, while the population also was growing at a high rate of 3.5 percent. Thus, there was a 2.8 percent annual increase in per capita output at the height of the dynamic period.[7]

The beginnings of the general transformation described by Kuznets as characteristic of the change from a traditional to a modern economy, as well as the elements of preconditioning for the take-off into self-sustaining growth, as defined by W. W. Rostow, became well marked in Argentina during this period.[8]

In contrast to the commonly recognized dynamic period, subsequent history has been subject to widely differing interpretations. Alejandro Bunge, who first applied the term *stagnation* to Argentina, believed that the economy began to stagnate when railroad building came to an end in 1914.[9] Although Alvin Hansen's stagnation thesis regarding the roles of the closing land frontier, declining population growth, and the cessation of massive innovations (such as the railroads) in the United States did not emerge until the 1930s, in retrospect it seems to have some application to Argentina.[10]

Díaz Alejandro dates the beginning of stagnation to 1930, when a severe foreign exchange bottleneck occurred as a result of the world depression. From 1925 to 1965, the overall rate of growth in gross domestic product averaged 2.7 percent per year, and the per capita growth rate was only 0.8 percent.[11] The rural sector (agriculture and livestock), which had expanded at an annual rate of 3.5 percent during the first three decades of the century, grew at only slightly more than 1 percent after 1930.[12]

Rostow, on the other hand, considered that Argentina had entered the take-off to self-sustaining industrial growth during the 1930s. His students, Guido di Tella and Manuel Zymelman, initially accepted this view, but later concluded that the apparent acceleration of the 1940s was arrested by the domestic economic crisis of 1952, when a high rate of investment began to yield relatively low increments of output.[13] Aldo Ferrer selects 1948, during the first regime of Juan Domingo Perón, as the beginning of stagnation, associating it with a decline in per capita product and the beginning of chronic inflation.[14]

Raúl Prebisch and his ECLA team regarded the period 1954–1957, when per capita output leveled off, as the onset of stagnation, but their investigation revealed evidences of sluggish growth and specific sectorial lags that antedated the Perón period.[15] Other analysts have applied the concept of stagnation to the recent period of "stop-go cycles," treating them as a technical trap from which a succession of governments has been unable to lift Argentina to a new growth path.[16]

This review indicates that *stagnation* is a very imprecise term, but its persistent use suggests that the Argentine economy has long been subject to unsatisfactory performance in one aspect or another. In recent years, Argentina seems to have reached a plateau of development affording a moderately high standard of living yet only sporadic improvements in the quantity and variety of goods available for consumption. The forces of applied technology have had, and continue to have, a powerful effect in preventing a general deterioration of the economy; yet the influence of prevailing social institutions is to limit the effect of these forces, prevent the formation of a fully integrated industrial economy and create social tensions that erupt in periodic political crises. Indeed, it is doubtful, as some have believed, that Argentina as a diversified economy has ever entered a period of continuous self-sustaining growth. While some parts of the industrial sector, in a protected environment, continue to support growth, early achievements in agriculture and petroleum production seem to have lost their vitality. Other elements, such as the railroads, port system, and power and communications networks, have been allowed to deteriorate, thus undermining the productive substructure that had been laid down.

## Technological Fusion and Accelerated Growth

Why did the Argentine economy grow so rapidly from 1870 to 1914? The chief factor was the process of "technological fusion" represented by the introduction of a complex of technological innovations, engineering and managerial skills, and educational methods that were almost wholly novel to the prevailing Latin American culture. These changes were introduced into a geographic region that was both endowed with exceptional natural resources and singularly receptive to development by virtue of institutional and demographic circumstance.

The term *technological fusion* as used here is akin to Kuznets' concept of "epochal innovation"; it has been substituted for the term *technological combination,* adopted by Ayres, which does not seem to convey the important element of a novel outcome.[17] *Technological explosion* has also been used, but it is less satisfactory because it suggests a destructive process, while "fusion" may be the forerunner of constructive development.

*Technological fusion* is defined as the coming together of a sufficient number of new technical elements to form a critical mass, with a resultant dynamic effect on economic growth and development. Prominent earlier examples have been the invention of printing in the mid-fifteenth century, the synthesis of the oceangoing sailing ship later in the same century, the Industrial Revolution of the eighteenth century, and the mechanization of agriculture in the nineteenth and twentieth centuries.[18]

In the mid-nineteenth century, the technological basis of Argentine life remained quite primitive. The indigenous cultures had left few traces, and the interior economy of the *pampas* depended on the most rudimentary of technological devices—the *gaucho's* saddlehorse, his knife, and his *boleadoras,* a rawhide hunting snare.[19] The remaining technological complement—oxcarts, sugar mills, wine presses, and handlooms—were derived mainly from the Spanish motherland in the colonial period or, in some cases, from indigenous sources. Few productive innovations were added after the colonial period, and the rural population remained largely illiterate and custom-bound. There was little foreign trade except in mules, salted hides, and jerked beef, and Buenos Aires was a small, relatively insignificant port whose estuary oceangoing ships could enter only with difficulty. Internally, the country was beset by continual warfare among rival *gaucho* chieftains (*caudillos*) and dwindling indigenous tribes.

The principal cultural and technological innovations stimulating the Argentine transformation entered the country from abroad in a succession of waves primarily affecting the agricultural, commercial, and transport sectors. English breeds of sheep imported by immigrant ranchers from the British Isles began to replace the poor, native flocks and made wool the major export product in the 1870s and 1880s. This stage was followed by the upgrading of cattle, also with English breeding stock, the installation of cheap barbed wire fencing and steel windmills of British or North American design, and the introduction of clover and alfalfa, which increased the carrying capacity of the Argentine cattle ranges and permitted meat exports to take the lead.[20]

At the same time, the construction of railroads, port facilities, and packing houses utilizing European engineering techniques transformed the Littoral of the Rio de la Plata and the Pampa into an integrated production zone highly complementary to the British economy. Argentina acquired a railway network (the first in Latin America) larger than that of the British Isles during this period, but as Bunge complained, its effects were largely confined to the agricultural and commercial sectors. Few market towns and regional industries came into being.

The key innovations of the period—methods of processing and shipping meat long distances under refrigeration—were the results of extensive experimentation in France, England, the United States, and Australia, with many attendant failures.[21] The discovery that beef could be chilled, rather than frozen, and shipped across the tropics to arrive in England in good condition opened up a vast new Argentine market for a preferred (price and income-elastic) good. This stimulated the construction of *frigoríficos* (packing houses equipped with mechanical refrigeration) and the technical perfection of efficient refrigerated steamships. By 1905 Argentina had displaced the United States as the chief exporter of fresh beef and mutton to the British market.

The cereals phase of agricultural development, dominated by wheat production, followed. By World War I, wheat-growing became increasingly mechanized with imported machinery. Grain exports were facilitated by the application of Dutch techniques of dredging and diking in developing the great grain port of Rosario and in improving the artificial port of Buenos Aires. Urban settlement of the Littoral and the consequent stimulation of construction, urban transport (including the building of a subway system in Buenos Aires), and the use of thermal electric power fed the growth process. Little, however, was done to create an infrastructure for basic industry, except for the forward-looking program of Domingo Faustino Sarmiento, who was president of the republic from 1868 to 1874 and later, national director of schools.

Sarmiento had an exceptionally keen insight into the development needs of his country. He encouraged railway construction and free immigration, but his most significant achievements were the establishment of a national system of popular education and the creation of new institutes of scientific investigation.[22] The recruitment of sixty-five young North American women to establish normal schools throughout the interior, initiated by Mary Mann (the widow of Horace Mann), was to have a profound effect on Argentine literacy. Both school facilities and enrollment almost doubled during Sarmiento's term, reaching 1,645 schools and over 100,000 pupils.[23] School attendance was made compulsory through the age of fourteen; although this rule was often violated, national illiteracy fell from more than two-thirds in the census of 1869 to one-third by 1914.[24] Improvements in the structure of the universities placed Argentina clearly in the lead in Latin American higher education; as a consequence, Buenos Aires became the region's cultural center in such fields as newspaper and book publishing, serious music, and stage and motion picture arts.

Institutional circumstances that facilitated the early growth process were the political unification of the country and the establishment of a national fiscal system by President Bartolomé Mitre in 1862. The customs receipts of Buenos Aires at last became the revenue of the nation, interior trade barriers were reduced, and a national monetary and postal system gradually emerged. These organizational changes laid the institutional basis for an integrated national economy for the first time.

Ayres has emphasized the role of the frontier in loosening the hold of encrusted institutions and permitting cultural cross-fertilization to take place.[25] In Argentina, the existence of a frontier culture had both positive and negative effects. The country escaped a neocolonial plantation economy mainly because its products, except for cane sugar and cotton in the north, neither required nor justified the use of masses of resident farm hands. Slavery, being unprofitable, disappeared, and with it the potential problems of racial dualism that have affected many other Latin American countries as well as the United States. However, the absence of a public domain and the uncontrolled speculative distribution of land prevented the emergence of a class of yeoman farmers, and land holdings soon became heavily concentrated. Immigrant tenant farmers were permitted to grow field crops only on short three- to five-year rental contracts and on condition that they would leave the land in improved alfalfa pastures for conversion to cattle ranges when their contracts were up.

As the export trade grew, the availability of cheap return passage in empty cattle and grain ships fostered heavy immigration. The newly arriving colonists were unable to take up desirable land except as small tenants and thus joined the migratory labor force or settled in the cities of the Littoral. Some handicraft industries were introduced by these immigrants, but most of the new arrivals did not come from the industrial zones of western Europe. They became the packinghouse workers, small tradesmen, and domestic servants of a growing metropolis. Although the frontier provided opportunities for individual enterprise, free trade discouraged the growth of domestic manufactures.

The integration of Argentina with the British trading community gave the country the advantages of an established commercial system, access to European financial markets, and a stable, gold-based currency, but it also created a relationship of administrative dependency. A significant feature of nineteenth century international financial institutions was that they provided, through the limited liability of the joint stock company and legal bankruptcy, for the liquidation of unsuccessful ventures. The Argentine economy began to lose the advantages of this arrangement when the government accepted the responsibility for guaranteeing essentially private investment, thus converting business failures into a funded public debt burden. However, commercial and investment credit remained functionally separate during this period, so a temporary crisis in the balance of trade neither undermined the usefulness of long-term financing nor seriously interrupted access to the international capital market. In general, although income flowed heavily to the landed *estanciero* class, Argentine investors were willing to take few risks on new enterprises during the growth period, preferring to use their profits to acquire more land and mortgages. It was not for lack of local financial resources that industrial investments remained in foreign hands and under foreign managerial direction.

In summary, an examination of the promotive forces in Argentine growth during the dynamic phase reveals the crucial role of borrowed technology and the remarkably passive role played by *criollo* entrepreneurship throughout the period.

## Institutional Rigidity and Stagnation

Why did a chronic and widespread tendency toward economic stagnation and technological dependency set in after World War I in most parts of the Argentine economy? Two related factors seem to be chiefly accountable: the inability to domesticate and propagate a technological process derived from outside sources and the resumption of cultural attitudes and habits embodied in institutions which were not suited to the promotion of an ongoing, diversified economy.

Notwithstanding an auspicious beginning, the technological culture represented by innovations from abroad remained largely alien to Argentine behavior patterns, except as the new practices could be adopted by a few native *criollo* groups and by an increasing inflow of Europeans from areas where the culture was already familiar. Moreover, the utilization of the new practices remained (to a remarkable degree) imitative (as in Japan following the Meiji Restoration) and in few respects became truly innovative (as it later did in Japan). Argentine society never succeeded in wholly adapting the foreign technological process necessary for economic development to become an internally generated transformation. This deficiency became acute when the country was thrown increasingly upon its own resources after the world economic crisis of 1930 and particularly when, after 1946, the government attempted wholly autarkical development. An insufficient basis had been laid for the internal generation of the ongoing functional components of growth (as was done, by contrast, in the Soviet Union, Japan, and Israel).

The economic symbiosis of Argentina and Great Britain was dealt severe shocks by World War I and the Great Depression, but only the latter was sufficiently critical to lead Argentina to revise its policy of domestic development. Di Tella and Zymelman have identified the "Great Delay" from 1914 to 1933 that prevented Argentina from beginning a process of industrialization until the conditions for successful import substitution were perhaps at their least propitious, in the depths of the Great Depression.[26] In the meantime, however, the opportunity to lay an educational foundation and to develop the necessary technical leadership and skilled manpower for the transition to a new stage of development had been largely missed.

It has been argued that a dominant social class such as the Argentine landed oligarchy could hardly be expected to foresee its own demise and plan for its succession, yet the question goes beyond deliberate choice and conscious motivation. As Ayres so effectively demonstrated, it is a matter of social conditioning. H. S. Ferns, in *Britain and Argentina in the Nineteenth Century,* remarks that the history of nineteenth century Argentina "commences with men but it ends with processes."[27]

In a sufficiently fluid and evolving society, the emerging educational opportunities and pragmatic experiences may bring forth a process of social trial-and-error that will result in increasing technical sophistication and, thus, have a growth effect. In the case of Argentina, this possibility was only partly realized.[28]

The *estancieros* who came to dominate Argentine society during the early part of this century were direct products of *criollo* society, which meant, in most instances, the rudimentary *gaucho* culture. They had little appreciation for the significance of invention, discovery, and adaptation, nor for education as part of the process of industrial growth. However, as they became the beneficiaries of the export bonanza, they rapidly adopted the modes of conspicuous consumption inspired by contact with Victorian and Edwardian England. The wealthiest of them sent their sons—the notorious *niños bien* with their retainers and strings of polo ponies—to the Sorbonne and to Madrid rather than to foreign institutes of agronomy, veterinary science, and engineering. Their concept of education, as Osvaldo Sunkel has said, was ornamental rather than functional. (Thorstein Veblen might have said ceremonial rather than workmanlike.)

The vitality of the system of public education laid down by Sarmiento and continued by his energetic protegé and minister of education, Nicolás Avellaneda (who succeeded him as president), rested, in large part, on the experimental and socially relevant methods of education introduced by the sixty-five North American "daughters of Sarmiento." After Sarmiento's passing, however, the innovative stimulus seems to have declined, both as to method and as to content. Neither the contemporary functional education movement introduced by Enrique C. Rébsamen to the limited zone of Xalapa in Mexico nor the later educational revolution precipitated by John Dewey in the United States caught on in Argentina, with the result that rote instruction given in half-day classes became the prevailing mode of elementary education. The expenditure of funds was increasingly concentrated in Buenos Aires, and children of the lower class in the interior cities and towns rarely completed more than two years of instruction. Access to the secondary schools and universities became the privilege of the children of the landed elite and the urban middle class, and law and medicine, the routes to professional status and wealth, became the approved university fields, while agronomy, animal husbandry, and viniculture were taught only in secondary institutes.[29]

The newly founded scientific centers were poorly supported, except in such fields as natural history and medicine, and in the former area they continued to carry on routine taxonomic investigations while the fruits of Darwin's discoveries (some of them made in Argentina) were inspiring novel investigations in genetics and anthropology elsewhere. The attitude of wealthy Argentines seems to have been: Why invest in scientific research, with its uncertain benefits, when the most advanced technology of the world or its products can be bought with the sterling proceeds of cattle and wheat?

Even the continuance of this source of wealth began to be affected. Many landowners moved to townhouses in Buenos Aires and left their *estancias* in the charge of majordomos, to the neglect of soil conservation, local improvements, and interior communications. Only stock-breeding captured their interest and continued to advance.[30]

Military engineering, so important in the development of North American waterways and harbors and, hence, in the basic transport system, was chiefly confined to naval operations. When later called upon by President Perón to provide direction for mining and manufacturing, military technicians proved severely deficient in training.

The electoral reforms which permitted the Radical Party to come to power in 1916 provided an opportunity for a change of direction. However, the middle class *criollos* and working class immigrants who formed the basis of the party upheld essentially the same values as the *estanciero* oligarchy they had temporarily displaced.[31] Hipólito Yrigoyen, their mystical leader, had essentially no program for an economic transition (much less a take-off) and soon began to suppress the very workers who had helped to bring him to power. Until the crisis of 1930, Argentina continued the policies of the "Great Delay" and made little progress toward structural transformation.

Only the University Reform of 1918, which began in Córdoba, emerges as a positive change in the direction of providing the human capital needed for development during this period. Unprogressive professors were replaced, university government was made more representative, and access to higher education was widened. Yet the fields of pure and applied sciences, except for medicine, were neglected, and there are few indications that invention and discovery (or their later forms, research and development) were actively promoted, in the universities, public institutes, or business. As a consequence, and on the basis of available evidence, the technological innovations that Argentina has given to the world must be regarded as negligible.[32]

## Exogenous Shocks and the Failure to Take Off

In an illuminating analysis of the Puerto Rican growth experience, Luz Torruellas has pointed out that dependent economies sometimes receive severe exogenous shocks, which may take a negative or a positive form.[33] The Great Depression was a negative shock of great moment to Argentina. It drastically reduced the export market, which had taken the bulk of the production of the Pampa before 1929, and therewith slashed the country's normal inflow of equipment and preferred consumer goods. The Argentine government was forced to adopt a policy of import substitution at a time when the advanced countries were "exporting unemployment" as never before. Many "penetration industries" established during this period, although characteristically highly competitive and self-sustaining, required continuous protection and never became independently viable. Entrepreneurial history is deficient, yet it indicates that oligopolistic practices entered Argentina almost at the outset of the new era of industrialization.

World War II, which isolated Argentina from normal commercial contact with the advanced countries, permitted the managerial inefficiencies of the 1930s to become standard practice and provided the country with neither the incentive nor the capacity for the intense technological activity that the war engendered among the principal belligerents. As a study by ECLA later revealed, managerial inertia and plant obsolescence reduced the ordinarily dynamic character of petroleum and electrical energy production, rail and road transport, and commercial agriculture during this period, while a marked shift of the labor force toward less productive forms of employment began to occur.[34]

The end of World War II fortuitously provided Argentina with the only opportunity for a strong positive shock that it has had in this century. The rapid accumulation of exchange reserves, made possible by the resumption of demand for food in Europe and the existence of large, unsold stocks in Argentina, could have provided the "seed corn" for extensive development. In a much discussed model, President Juan Perón and his consort, Eva, made spectacular changes in social policy, but, despite their aspirations for an independently developing country that could be strongly influential throughout Latin America, they exemplified the very institutional behavior that had previously obstructed growth. Their errors of judgment with respect to development can only be suggested here. These included a systematic depletion of resources ("de-capitalization") in the agricultural sector, the use of large foreign balances to purchase a railway system already installed but in disrepair, the failure to admit imports of standby equipment for essential power production, the neglect of petroleum exploration and development, the mismanagement by military administrators of industrial plants that had been declared alien property, the commitment of public funds to monumental construction projects of little productive use, and the failure to extend the highway system or to provide motor transport as the railways fell into unreliable operation.

The effect of the Perón administration on manpower utilization and human capital formation is particularly noteworthy. The normal flow of population from rural to urban areas was accelerated to a degree that, together with the unavailability of machinery, forced farm operators to return to less labor-intensive methods of production.

Under similar circumstances in the United States, the loss of farm manpower facilitated mechanization, in both the wheat and the cotton regions. By contrast, Argentine landlords returned to cattle production or allowed their land to lie idle and grow up in thistles. Much of the manpower released from the land was absorbed in a pronounced overstaffing of the railroads and the administrative services. The government encouraged higher rates of consumption at the same time that the productive base was being eroded, and this policy was, no doubt, a contributing factor to the subsequent persistent inflation.

Most of the advantages of the University Reform of 1918 were undone after 1943. Under the doctrine of "*¡Alpargatas si, libros no!*" the most distinguished university professors were obliged to resign en masse (a procedure that has been repeated by subsequent governments), political indoctrination became obligatory, standards of admission were reduced, and examination procedures were corrupted.[35] The quality of a university degree was significantly lowered, and the conduct of technical studies became subject to political intervention. To its subsequent embarrassment, the Perón government invested heavily in an abortive project to develop the peaceful uses of atomic energy. It proved to be headed by a scientific charlatan.

At the elementary and secondary levels, instruction was also heavily politicized, and teachers were placed under the surveillance of their own pupils. Thus, the effects of educational intervention were transmitted to a new generation.

It is difficult to magnify the structural damage to the Argentine economy that occurred during the first Perón period, although its historical roots extend further back. An alternating succession of elected and military governments has since been unable, for the most part, to correct the critical sectorial lags. Some of the experience, however, has been instructive. In a sharp reversal of policy in 1959, President Arturo Frondizi allowed a group of foreign oil companies, with outside engineers and equipment, to apply their skills to the Argentine petroleum fields. The action was construed as an ideological betrayal and later reversed, yet the temporary outcome demonstrated conclusively that the problem of petroleum production was essentially technological and managerial and that Argentina had the capability for becoming self-sufficient in this resource.

## Cultural Interdependence and the Transfer of Technology

The foregoing analysis seeks to explain the sluggishness of important sectors of the Argentine economy—and the phenomenon of stagnation—as a result of institutional obstacles that prevented Argentina from availing itself of a fully domesticated technological process and that produced policies inappropriate to promote the accelerated growth required for a take-off. It is necessary, however, to account for the fact that, in spite of all its problems, Argentina remains the country in Latin America with the highest *general* standard of living (as distinguished from the highest average per capita income, attained by Venezuela). This is not merely a heritage from the past. Nor is it wholly attributable to a more equitable distribution of income.

The growth of Argentina in the last four decades has been quite lopsided. Some sectors have shown considerable dynamism, while others have lagged. It was particularly the growth of the manufacturing industry as a "leading sector" in the 1930s that led Rostow, Di Tella, and Zymelman to conclude that Argentina was approaching a take-off.

In a passage that might have been written about Argentina, Kuznets has explained in general terms that such a process may be misleading:

> At any given time in the history of a country's economy some dynamic "leading" sectors are the loci of dynamic growth which, through various linkages, induce growth elsewhere in the economy. But unless the relation between changes in these modern "leading" sectors and the rest of the economy is significant, stable, and general, a marked rise in these modern sectors may have little effect on the persistently stagnating remainder of the economy of some countries and thus on their overall growth.[36]

The progress in manufacturing and related productive sectors has not been sufficient to lift Argentina to an accelerated growth path, except for short intervals, yet it has enabled the economy to operate at a fairly sustained and moderately high level. This merits explanation. Díaz Alejandro has attempted to measure the differential contributions of major sectors to growth from the period 1927-1929 to the period 1963-1965 by comparing their net additions to gross domestic product measured at factor cost. Results vary considerably depending on whether 1937 or 1960 prices are used as a base, but, in both cases, manufacturing stands out as the major contribution to growth during this period.[37] Other major contributions are the transport and government service sectors; the rural, oil and mining, construction, and communications sectors lag notably. In part, the increases in transport and government services are explained by a relatively heavy flow of manpower to these sectors induced by policies of the earlier Perón regime and also by a statistical aberration. Díaz Alejandro points out that:

> the contribution of most services to Argentine output is measured by quantifying their *inputs* (e.g., government services essentially measure employment), without looking too closely at changes in the quality of these services. The high proportion in the increase of GDP between 1927-1929 and 1963-1965 accounted for by all services . . . plus the generalized impression that the quality of many services has deteriorated, strengthens doubts as to the extent of real growth during the last thirty-six years.[38]

Nevertheless, the growth of industry is more than a statistical illusion. It rests on the exceptional access to the international transfer of technology that characterizes this sector. At times, as during the Frondizi administration, import restrictions are relaxed, and a flood of new equipment enters the country as industrialists take advantage of an opportunity for which they have been waiting and often accumulating foreign bank balances. Even when the decision to allow capital goods imports is not wholly rational, as in permitting ten new firms simultaneously to begin producing automobiles for a protected market estimated to have an absorptive capacity of only 200,000 vehicles a year, the machine tools, once in the country, are adaptable to other uses.

International companies also find ways to supply their subsidiaries with new techniques, new product lines, and the necessary financial support. Private domestic manufacturers, through trade fairs, product licensing, franchise agreements, and similar sources, tend to have more contact with new products and processes than do state enterprises in the transportation, petroleum, electricity, and communications fields, or private agricultural enterprises dependent on deficient public research agencies. Job changes by managers and engineers who move from firm to firm have also produced innumerable side effects in transferring industrial know-how throughout the private sector. The "spread effects" are probably greater than heretofore suspected.

In a recent investigation, Jorge M. Katz found that while the gross domestic product of Argentina rose at an annual cumulative rate of 3.3 percent between 1960 and 1968, cumulative growth in the manufacturing sector reached 4.4 percent.[39] Within this sector, the 200 largest manufacturing plants in nine branches of industry had an annual growth rate for the same period of 9 percent, or double the annual growth rate of the industrial sector as a whole. The plants with the highest rates of growth in output were those with the highest rates of growth in capital stock per worker and those with the highest rates of technical progress, as measured by overall factor productivity. The most dynamic firms in this regard were concentrated in the metallurgical, electrical, and chemical industries—those with strong links to foreign sources of technology.

Much of the recent discussion of the role of the multinational corporation in dominating and controlling the spread of advanced technology deplores the resulting "dependence" of the less developed countries—who must receive their innovations and improved techniques in this way, often at costs that seem unconscionably high. This complaint overlooks the fact that, in the short run, such countries really have no alternative if they wish to gain access to the rapidly changing international stock of useful knowledge. Having developed no internal sources of technology, these countries must depend on outside sources.

Moreover, contrary to Katz's conclusion, the evidence of increased productivity suggests that however expensive the acquired techniques may seem, they pay off. Argentine users of borrowed technology may consider that they are in a poor bargaining position, but the results more than warrant the costs, when there are no domestic alternatives.

In actuality, the nature of the international technological transfer process is such that complete national independence, amounting to a form of cultural autarky, is no longer possible or desirable (as leaders of the Soviet Union have come to realize). Rather, a modern country must seek a form of technological *interdependence* in which it can be an active contributor as well as an effective recipient of new techniques as they become attainable. This is the only meaning that *technological independence* can have for latecomers.

## The Prospects for Developmental Autonomy

Although it is not yet possible to give a secure empirical foundation to the conclusions advanced in this chapter, if it is indeed true that the basic causes of the extended Argentine stagnation problem are deeply cultural and involve a failure to internalize the technological modes of behavior common to the advanced countries, serious policy implications are raised. The inference may be drawn that it will require a profound cultural reorientation to restore the earlier growth conditions. Also, a new strategy of development would have to systematically seek out the deficiencies in present practices and the means of repairing them. This cannot be accomplished overnight, but the reversal of policy must be made if Argentina is to have any hope of overcoming technological dependency and exercising greater autonomy over its own developmental process.

Moreover, with the passage of time, productive processes and international market relationships become more complicated and sophisticated, and the race to catch up in technique becomes more difficult. Individual invention no longer suffices, and organized research and development become essential. Meanwhile, old habit patterns become ingrained and political frustrations mount.

It is, therefore, essential and urgent that a national development strategy include systematic attention to the reformation of educational institutions from the nursery school to the most advanced research institute in order to foster a problem-solving environment. These formal institutions and their ancillaries—industrial apprenticeship systems, farm extension services, and other forms of extension education—are vital to foster the growth process.

This is not to say that educational reform is a sufficient condition to restore Argentina to a growth path. The economy is too complex and its current problems too intense for so simple a solution. However, a genuine educational reform implies a radical change in attitude on the part of governmental leaders, industrialists, and even professional educators. Education must no longer be regarded as essentially a form of political activity, but rather as a necessary prerequisite to the effective use of modern science and technology. To accomplish this shift, it is necessary to create a new atmosphere of vigorous and free inquiry. No one understood this better than Sarmiento, and Argentines might well take a leaf from their own history.

## Notes

1. The principal analyses of the Argentine stagnation problem have been made by Raúl Prebisch and a team of the United Nations Economic

Commission for Latin America (ECLA), *El desarrollo económico de la argentina,* mimeographed (Santiago, Chile: ECLA, 1957–1958), and *Análisis y proyecciones del desarrollo económico. V. El desarrollo económico de la argentina* (Mexico, D.F.: U.N. Department of Economic and Social Affairs, 1959); Aldo Ferrer, *The Argentine Economy* (Berkeley: University of California Press, 1967); Guido Di Tella and Manuel Zymelman, *Las etapas del desarrollo económico argentino* (Buenos Aires: Editorial Universitaria de Buenos Aires, 1977); and Carlos F. Díaz Alejandro, *Essays on the Economic History of the Argentine Republic* (New Haven: Yale University Press, 1970). The profession is enormously indebted to Professor Díaz Alejandro for assembling and refining the basic data for the Argentine economy to 1965.

2. Francisco C. Sercovich, "Dependencia technológica en la industria argentina," *Desarrollo Económico* (Buenos Aires) 14 (April-June 1974):33–67; and Jorge M. Katz, "Industrial Growth, Royalty Payments and Local Expenditures on Research and Development" in *Latin America in the International Economy,* ed. Victor L. Urquidi and Rosemary Thorp (London: Macmillan, 1973), pp. 197–224.

3. C. E. Ayres, *The Theory of Economic Progress,* 2nd ed. (New York: Schocken Books, 1962). See especially ch. 5–11.

4. Simon Kuznets, "Modern Economic Growth: Findings and Reflections," *American Economic Review* 63 (June 1973):247.

5. See especially Kuznets, *Modern Economic Growth: Rate, Structure and Spread* (New Haven: Yale University Press, 1966).

6. Díaz Alejandro, *Essays,* pp. 2–3.

7. ECLA, *Análisis y proyecciones,* pp. 4, 400. Cited by Díaz Alejandro, *Essays,* pp. 6, 8.

8. Kuznets, "Notes on the Take-off," in *The Economics of the Take-off into Sustained Growth,* ed. W. W. Rostow (London: Macmillan, 1963), pp. 22–43; and W. W. Rostow, *The Stages of Economic Growth* (Cambridge: University Press, 1960), pp. 17–35.

9. Alejandro E. Bunge, *Las Industrias del Norte,* vol. 1 (Buenos Aires, 1922), p. 159. Cf. Díaz Alejandro, *Essays,* p. 53n.

10. Alvin Hansen, *Full Recovery or Stagnation?* (New York: W. W. Norton, 1938), pp. 279–80, 288–89.

11. Díaz Alejandro, *Essays,* pp. 67–69.

12. Ibid., p. 71.

13. Guido Di Tella and Manuel Zymelman, *Las etapas,* pp. 22–32, 142.

14. Ferrer, *Argentine Economy,* pp. 174–75, 185. The average annual

rise in the cost of living in Buenos Aires reached 39 percent during the period 1955-1959 and 23 percent in 1960-1964. Díaz Alejandro, *Essays,* p. 365.

15. ECLA, *Análisis y proyecciones,* p. 3 and passim.

16. Oscar Braun and Leonard Joy, "A Model of Economic Stagnation—A Case Study of the Argentine Economy," *Economic Journal* 78 (December 1968):868-87; and Reinaldo F. Bajraj, "Some Notes on the Argentinian Economy," Paper presented at the University of Cambridge, England, 14 December 1972.

17. Ayres, *Theory of Economic Progress,* pp. 112-24. "An epochal innovation may be described as a major addition to the stock of human knowledge which provides a potential for sustained economic growth—an addition so major that its exploitation and utilization absorb the energies of human societies and dominate their growth for a period long enough to constitute an epoch in economic history." Kuznets, *Modern Economic Growth,* p. 2.

18. Ayres, *Theory of Economic Progress,* pp. 127-52; Kuznets, *Modern Economic Growth,* pp. 1-16. See also James H. Street, *The New Revolution in the Cotton Economy: Mechanization and its Consequences* (Chapel Hill: University of North Carolina Press, 1957), pp. 91-156.

19. James R. Scobie, *Argentina, A City and a Nation* (New York: Oxford University Press, 1964), pp. 67-68.

20. Simon G. Hanson, *Argentine Meat and the British Market* (Stanford: Stanford University Press, 1938), pp. 11-16, 100-101, 117-18. The British breeds of sheep and cattle must be regarded as "technological" innovations, since they represented the products of generations of controlled breeding, a process completely unknown to the *gaucho.*

21. Ibid., pp. 18-47.

22. Hubert Herring, *A History of Latin America from the Beginnings to the Present,* 2nd rev. ed. (New York: Alfred A. Knopf, 1961), pp. 650-55. See also Alice Houston Luiggi, *65 Valiants* (Gainesville: University of Florida Press, 1965).

23. Herring, *History of Latin America,* p. 654.

24. Arthur P. Whitaker, *Argentina* (Englewood Cliffs: Prentice Hall, 1964), p. 41.

25. Ayres, *Theory of Economic Progress,* pp. 133-54.

26. Di Tella and Zymelman, *Las etapas,* pp. 71-101.

27. Henry S. Ferns, *Britain and Argentina in the Nineteenth Century* (Oxford: Oxford University Press, 1960), p. x.

28. The indispensable nature of scientific and technical education in advancing economic development is emphasized by both Ayres in *The*

*Theory of Economic Progress,* p. xxi and passim and Kuznets in *Modern Economic Growth,* pp. 289–93.

29. James R. Scobie, *Buenos Aires: Plaza to Suburb, 1870–1910* (New York: Oxford University Press, 1974), pp. 222–24.

30. The long neglect of public research and extension services in agriculture was relieved, and then only partially, by the creation of the Instituto Nacional de Tecnología Agropecuaria (INTA) in 1956. Díaz Alejandro, *Essays,* p. 190.

31. David Rock, *Politics in Argentina, 1890–1930: The Rise and Fall of Radicalism* (London: Cambridge University Press, 1975), pp. 265–74.

32. It is not, of course, necessary for an advancing industrial country to generate all of its own technology, as O. J. Firestone has shown for the case of Canada, where only about 5 percent of the patents currently in use are of domestic origin. See his *Economic Implications of Patents* (Ottowa: University of Ottowa Press, 1971), especially chapter 7. However, in the long run, a considerable price is paid by countries that do not enter the stream of world innovative practice as active participants.

33. Luz M. Torruellas, "Some Theoretical Insights into Puerto Rico's Recent Economic Growth," Paper presented at Rutgers University on 30 November 1962.

34. ECLA, *Análisis y proyecciones,* pp. 19–39.

35. The slogan translates loosely as "Up with the workers, down with the intellectuals!"

36. Kuznets, *Modern Economic Growth,* p.25.

37. Díaz Alejandro, *Essays,* p. 73.

38. Ibid., pp. 73–74, his italics.

39. Jorge M. Katz, "Industrial Growth," pp. 203–4.

# 13
# The Development of Alternative Construction Technologies in Latin America

*W. Paul Strassmann*

With a rapidly growing population, Latin America faces an acute shortage of housing that worsens with every passing year. A number of efforts have been made to relieve this problem by introducing foreign techniques of mass-produced housing, chiefly from Europe and the United States, in the belief that they would materially reduce the cost of construction and increase the available supply of new and more adequate housing. This chapter will review briefly the record of prefabricated industrial systems building (ISB) in two areas of Latin America—Puerto Rico and Colombia—and will consider possible alternative techniques that may have a better chance of success.

In politics, an old adage is that "you can't beat somebody with nobody." In science, theories are not destroyed by debunking; a different theory must replace an old one. So it is with construction technology. The appeal of prefabricated industrialized systems building is immense, and its widespread failures do not hamper new attempts at introduction. To point out that ISB requires an unduly large number of dwelling units at a site, greatly underpriced capital, extraordinary managerial skill, and unreasonable luck will not offset its hypnotic appeal. It is necessary to provide more practical alternatives.

## ISB in Puerto Rico

Industrialized systems building, or ISB, is the disciplined and integrated application of mass-production technology to construction and to the manufacture of large building components. Under scientific factory conditions, standardized components come off the assembly line ready for quick mechanical, usually dry, installation. Such methods have been reasonably successful in Europe, but have failed for housing in the United States, despite strong government support in the form of "Operation Breakthrough."

In the Commonwealth of Puerto Rico, however, ISB was a partial success, at least in one case. Compared with the mainland, living in high-density standardized apartment blocks in San Juan has proved both tolerable and necessary. Compared with the rest of Latin America, construction wages in Puerto Rico have been much higher, and capital has been much cheaper. Puerto Rico is more like Europe than the rest of the hemisphere.

Nevertheless, even Puerto Rico had ISB failures. In the 1950s, the International Basic Economic Corporation (IBEC) tried to introduce a megaform designed by Wallace Harrison. An entire two-bedroom house could be cast using this form, and a crane would install it, moving from lot to lot. The attempt was abandoned after a while and later tried in Chile, again without success.

Another failure was that of Shelley Enterprises. Their system involved open-ended boxes stacked in checkerboard fashion so half the enclosures were obtained "free." For stability the whole was post-tensioned with steel cables passing through the corners of all boxes from one story to another. Ten to 25 percent savings over previous construction costs were expected, but in a highly capital-intensive operation with inexorable fixed payments falling due, unforeseen delays proved fatal.

Eventually IBEC combined with the Rexach Construction Company and Larsen and Nielsen, a Danish firm, for another try under the name of RELBEC.[1] Investment in a heavy panel plant required $6.4 million, and only 125 workers were employed in eight-hour shifts to make the components for 1,500 apartments per year. Transportation costs limited the market to a radius of 50 kilometers, and yet if the firm had utilized two shifts, it could have supplied one-third of the housing being built in the San Juan area.

Meanwhile, the Puerto Rican Urban Renewal and Housing Administration attempted to introduce the French Estiot system. This system was estimated to be viable with an annual rate of construction of 300 units if it could be sustained for five years. At this rate, the fixed cost per dwelling would be $1,426, and the variable cost, $6,810, at 1970 prices. If the volume of construction fell to only forty-eight units (or two buildings) per year, the fixed cost would rise to $8,913. Dwelling unit costs would rise from $8,200 to $15,700, both far beyond the ability to pay of most home buyers in the rest of Latin America. Indeed, if labor costs were reduced to one-quarter the Puerto Rican hourly rates and if finance costs were doubled, the price would rise even further to $16,700 per unit.[2] Thus, the feasibility of this sytem was confined to the special conditions existing in Puerto Rico.

**ISB in Colombia**

Although ISB was barely viable under Puerto Rican conditions, its promoters in much less faborable Latin American settings were by no means discouraged. The United Nations and the Government of Denmark sponsored a Seminar on Prefabrication of Houses for Latin America in Copenhagen in August 1967. The following types were recommended:

a) Prefabricated houses having concrete panels for walls and 7.5 cm.-thick shell roofs, produced and assembled by the vacuum-concrete process;
b) Lightweight prefabricated concrete panels used within modular systems (extruded panels, partially finished framed panels, etc.);
c) Houses made with prestressed concrete prefabricated columns, beams, and slabs;
d) Multistory apartment buildings that use a heavy prefabrication system, where the maximum size and weight of the panels are 11.44 x 2.60 meters and 7 metric tons, and the bathroom block weight is limited to 6 metric tons;
e) Prefabricated houses transported and assembled as room units fully completed in factory. The units are connected together on site as two-room, four-room, and six-room houses and can be finished for occupation in a mean construction time of eight days;
f) Prefabricated metal frame houses with concrete sandwich panels, filled with treated wood-chip insulation for partitions and facades, supplied with glazing and prefitted plumbing in walls;
g) Partial concrete prefabrication and modular coordination in housing construction by aided self-help.[3]

Following the last recommendation, the Instituto de Crédito Territorial (ICT), the leading housing authority in Colombia, made six distinct attempts to introduce ISB into Colombian construction. The institute obtained capital to build three prefabrication plants in as many cities, in order to produce heavy load-bearing panels. Prefabricated elements included lintels, beams, door and window frames, staircases, and floor slabs. ICT had previously introduced other reinforced cement components for low-cost housing in 1965. The prefabricated elements were intended for self-help construction, and two men could handle any of fifty components to build two-story concrete block houses.

At the first Latin American Seminar on Prefabrication, held in Bogotá in April 1971, Santiago Luque Torres, an architect, reported that the most successful prefabricated products were rather conventional small components like bricks, blocks, tiles, window frames, basins, and pipes. More unusual were reinforced ceiling beams, sometimes made of lightweight cement and involving pre- or post-tensioning. Since they were not visible, their quality was intended to be structural rather than esthetic.

All of the efforts of ICT in introducing ISB into public housing construction ultimately failed. The quality of construction was too low, the volume too small, and the costs too high.

Other types of prefabrication with larger and more visible components gained success only under the auspices of large integrated enterprises organized for high-quality construction. The extra capital, managerial talent, and technical skill needed could only be found in such firms. They chose advanced prefabricated components, not to save resources, but to gain time in finishing expensive hotels, office blocks, or luxury apartments. For low-cost housing, their volume requirements were too high.[4]

A borderline case is the Outinord System. This system uses prefabricated metal formwork in the shape of inverted U's or "tunnels." Walls and ceilings are poured simultaneously. In multistory building, the builder can start an upper story before the lower one is completely set, since the forms provide support. Capital is substituted for about one-third of the site workers, and construction time falls by half. Nevertheless, the Outinord system is probably the most labor-intensive of modern European systems.

The system was developed in France around 1952 and introduced in Latin America some ten years later, with varying results. In Puerto Rico, where relatively high income levels made large-scale construction possible, it was a success. Elsewhere, those who obtained the patent rights and had forms built were disappointed. One Colombian contractor, who had acquired the Outinord system, reported in 1970 that a 250-unit contract was sufficiently profitable to permit adoption of the system. Three years later he acknowledged that he had been mistaken. A thousand units, enough work to last four years, he then believed was the minimum. Yet the government seemed to consider it politically unwise to commit so much work to any one contractor. For a builder, the equipment is so expensive that it cannot be treated as a sunk cost, easy to use off and on. Assembling and training a crew, mobilizing a crane, and setting up a site all add heavy expenses.

A contractor who has used the system successfully and continues to use it explained in a personal interview:

> If anything, the traditional system is just a shade cheaper, but Outinord is much faster. We put up a 20-story building in just 60 days, and I mean calendar days. Any traditional building that takes 100 days to put up, we can put up in 30. Not only that, but we use only 30 percent as much labor. Only a few of the most skilled workers, crane operators and the like, are on our permanent payroll. And remember, no finish is needed with our system. That helps offset the fact that the cement we use is more expensive than bricks.
>
> Of course, the system depends on having a minimum volume. That's why so many of the other people have failed with their crazy schemes. I'd say 15,000 to 20,000 square meters is the minimum. At 100 square meters per unit that's 150 to 200 apartments. For smaller ones it has to be a minimum of 200 units. Actually, we're not really tempted until it gets to 400 100-square meter apartments. When the government asked us to build 64 small apartments, we refused.
>
> It's not that we can't afford to have our equipment idle. What is critical is moving all the material to the site, especially the crane, and then not having it fully occupied. The system is definitely impossible for individual houses because of this. About half the time, we use the traditional system. In addition to Outinord we use prefabricated beams, not just for support of the ceiling, but for the structure itself.

A housing design using the Outinord system won a low-cost housing competition conducted at the United Nations *Proyecto Experimental de Vivienda* (PREVI) in Lima, Peru, in 1969. Actual attempts to use the system, however, showed that construction costs were at least $6.80 per square meter, at current U.S. prices, rather than $3.60, as claimed.

At the Bogotá seminar in 1971, an Argentine architect warned that building systems from industrialized countries might not be suitable for Latin America:

> Prefabrication, or better, the industrialization of construction, should rise in direct proportion to the degree of development. It may grow gradually from prefabrication of small components, which needs only a small investment in equipment and therefore allows sensitive adjustments to market fluctuations, to total prefabrication, which takes great capital expenditures and therefore a reliable market for reaching an optimal yield.[5]

Despite misgivings about the feasibility of prefabrication, the resolutions adopted by the 115 architects, builders, and materials producers attending the seminar supported government subsidies for more experiments with ISB. Governments were asked to finance prefabrication research, set up standards, train workers and professionals, and provide cheaper credit for ISB builders and home buyers. The group concluded that preference should not go to heavy, integrated, inflexible systems from abroad.[6]

## Non-ISB Innovations for Roofs and Walls

Novel alternatives to ISB *do* exist. Often they are not alternatives but complements to conventional building methods, and perhaps that is why they do not seem sufficiently forward-looking to editors of architectural magazines. But the most visionary idea, like the most conventional, can only be a point of departure for practical application.

Both the IBEC megaform in Puerto Rico and the Outinord system involve pouring cement on metal forms. As a cheap substitute for the forms, some architects have designed fabric balloons. These are blown up, and cement is sprayed on them. When the cement hardens, the balloon can be collapsed for use elsewhere. The Israelis have already built hundreds of dwellings in the Sinai desert with such a system, patented by Haim Heifetz. In 1969 the cost per square foot covered by such a shell was U.S. $3.10 for 33-foot diameters and U.S. $6.35 for 99-foot diameters.

But will a man's dome feel like his castle? In Latin America the answer seems to be "no." In Mexico, houses with prestressed domes could be rented, but did not have enough appeal to be sold. Nevertheless, the principle can be applied in modified form. One can stretch hessian cloth or burlap over a 30-inch square frame and pour on one inch of concrete. Its weight makes the cloth sag, creating an inverse domed tile weighing 50 to 60 pounds. When laid across a grid of concrete T-bars, these tiles make a strong capital-, materials-, and labor-saving roof. This system, borrowed from Pakistan by Ernesto Paredes, a Peruvian architect, saved $0.70 per square meter (U.S.), about 12 percent compared with conventional construction costs.

Nothing in housing is so disappointing for the occupants as a roof that leaks, burns, or collapses. Yet no housing component is more difficult to install or, consequently, more expensive. Roofs are a natural focus for innovation. They should be produced out of local materials that are easily cut, mixed, treated, poured, or assembled. They should last some two decades without repairs, absorbing wind but not rain or (in the tropics) solar radiation.[7]

An already widely used innovation for roofs is the long, interlocking, N-shaped, asbestos cement channel, invented by Álvaro Ortega of the United Nations, and called *canaletas* or *canalones*. No supports are needed, since a single piece goes wall-to-wall. A disadvantage is that this ceiling cannot also serve as a floor for an upper story. The price is difficult to estimate, since asbestos cement products are often made by a local monopolist who can vary their price up or down by 50 percent. The channels can be shipped in a compact way by stacking. Installation is merely a matter of setting them in place.

Other innovations which may serve as alternatives to ISB and complement conventional technology apply to walls. Some early work was carried out in the West Indies. A material called Megcrete, formed into blocks or panels, was developed in Barbados in the 1950s. Megcrete is composed of six parts of pressed bagasse, the by-product from sugarcane, and one part freshly slaked lime. Use of this material was reported to reduce the cost of houses, in local currency, to 12 shillings per square foot, compared with 14 shillings for timber houses built by the owners (18 shillings if government-built) or 15 shillings for a coral limestone house.[8]

Fired bagasse-clay bricks, a similar material, were developed in Antigua in the West Indies. This mixture of one part clay and one part bagasse (10:1 by weight) forms genuine bricks. Since the material is 20 percent lighter than clay bricks, some roofing tiles have been made as well, but these tiles have the disadvantage of being porous. Some of the bagasse-clay bricks can be made from inferior clay without cracking, but their shape is too poor for use anywhere except for invisible interior partitions. This attempt at innovation may be classified as interesting but inconclusive.[9]

In Peru, bitumen-stabilized adobe bricks are being developed by local factories in association with the International Institute of Housing Technology at Fresno State College, California. The Peruvian project uses only 1.5 percent bitumen, compared with 5 percent in California. Holes are left in the blocks for bamboo reinforcing against earthquakes. After the bamboo is inserted, the holes are filled with the same stabilized adobe. Hexagonal floor tiles that can be polished have also been developed. The cost of a house was estimated in 1973 at 1,000 soles per square meter (U.S. $2.30), or 40 percent of the conventional cost.

Besides bricks, walls consist of mortar, so one innovation is to eliminate the mortar. A way of doing it with interlocking blocks was worked out by the firm Educational Design, Inc. and has been demonstrated at PREVI in Lima. These interlocking blocks are known as the EDI Thermond system and weigh 5 kilograms. Reinforced concrete is needed only at the corners and upper edge of a wall; up to three stories, the walls are self-supporting. The blocks can be made by a single worker using a hand mold at the rate of 400 per day. The system has been used for low-cost housing in the southern United States and Mexico and is estimated to save 15 percent at wage rates prevalent in Peru.

## Alternative Brick Production Methods and Plastics

Modern brick factories cannot undersell sun-dried bricks made by traditional methods in poor countries. Typically, a box with two brick-sized compartments is filled with clay, which is pushed in by hand and leveled off with a stick. When the box is inverted, the bricks are left to cure in the sun. A diligent worker can make 800 to 1,000 bricks per day. After sun-drying, the bricks are fired for a week in crude trapezoidal ovens. The cost per brick is ordinarily one-third less than that of average common brick in the United States.

Notwithstanding the prevalence of sun-dried brick, modern brick plants first appeared in Lima around 1960, and their market share stabilized at around 30 to 40 percent. They produced the superior product required by modern construction. Precision cutting with wires gave the bricks more accurate dimensions, and more careful mixing and firing produced greater strength. Consequently, these bricks could sell for 12 percent more (2.4 U.S. cents) than hand-made bricks in 1968.

A category of innovation that has become significant is the hand-operated block or brick machine. The best known of these is the CINVA ram machine designed by a Chilean engineer for the Inter-American Housing Center in Bogotá in 1957. The machine is small, weighs only 140 pounds, and can be operated by one man. Its capacity when operated by two men is about 300 soil-cement blocks per day, using a mixture of 5 percent cement and 95 percent clay soil.[10]

In the West Indies, 200 pilot houses were built with CINVA ram blocks in 1964, and these were evaluated six years later. Some had eroded, but other were satisfactory. A variety of recommendations emerged: The silt content of the soil should be less than 10 percent. The cement content should be confined to the range of 5 to 10 percent. The minimum thickness of the bricks should be 6 inches, and blocks with over 8 percent water absorption should be waterproofed. They should not be used for foundations in rainy areas, and a damp-proof course should be laid at least 3 inches above ground level.

In recent years, proposals have been made to use plastics in housing construction, but this development is in its early stages and may not prove practical. A Canadian expert concluded that, "The inherent ideal properties of plastics suggest that by themselves they cannot form ideal house structures, now or for some time to come."[11] The British Building Research Station agreed:

> What are called all-plastic houses for developing countries claim advantages many of which seem to be very questionable . . . they suffer from lack of thermal capacity and high cost. . . . For normal everyday uses, complete buildings in plastics are likely to remain in the realm of fantasy.[12]

Nevertheless, plastics, if fireproof, can serve well as flooring, insulation, and pipes. Here we shall mention only the bamboo-polyurethane beam of Peru and the foam matrix roofing system of Washington University, St. Louis. The beams were developed by Christopher Alexander of the Center for Environmental Structure for the United Nations PREVI program. They are made of 6-centimeter bamboo rods placed over plywood templates, with a core of polyurethane fire-retardant foam, foamed in place. They are produced in standard dimensions of 20 x 40 centimeters, 5 meters long. They cost half as much as a comparable reinforced concrete beam and weigh only 40 percent as much. They can be cut with simple tools and handled by two men. If the foam is available cheaply within a developing country, it provides a means to use other local materials economically.

The low-density foam matrix reinforced with local reinforcing fibres used at the Center for Development Technology of Washington University is receiving field trials in three climatic zones of Mexico.[13]

**Conclusion**

The most that can be expected from the best prefabricated building system is that it will reach the cost range of conventionally built housing: around $40 per square meter in 1970 dollars. For this goal, the system will need a five-year volume of 1,500 to 2,000 dwelling units in four- to five-story apartments at one site. One hundred man-years, or 3 percent, of the on-site jobs will be displaced.

The vast majority of systems need a much larger volume, far exceed conventional costs, and displace hundreds of additional workers. Two decades of failures with such systems in Latin America and Africa have been witnessed, yet, as new administrations come to office, ISB again and again is given another try. Left- and right-wing architects alike are fascinated by anything systematic and neat that seems to replace chaos.

Speed in dwelling unit construction is readily confused with lower cost and with the annual rate of building that a country can sustain. Industrial systems building promoters rarely inform inexperienced military or political housing authorities about start-up delays, on-site interruptions, breakage rates, diseconomies of small scale, and other problems that arise in practice. Besides, no one can deny that modular coordination and prefabrication are desirable for certain components.

This discussion has suggested that promoting alternative technologies may be a way to cope with the allure of ISB. Of dozens of possibilities, only a few relating to roofs and walls have been cited.[14] Such examples of unquantified cases cannot, of course, be conclusive, but they suggest that alternatives exist and should be explored further before large investments are made in systems of doubtful value.

Some economic and social issues are beyond solution with better physical implements; one cannot solve nontechnological problems with technology, especially not the deficiencies of ISB. With a farsighted and dispassionate policy toward land and mortgage finance, resources can be made to flow into residential construction using a rationalized conventional technology and with a minimum of waste. As long as unemployment is severe, measures such as withholding tax deductibility from heavy prefabricating equipment are sensible.[15] A rational policy for housing in the less developed countries should take into account effective use of local materials, utilization of abundant labor, and reduction of costs in order to bring a greater supply of acceptable dwellings to the growing populations for whom more glamorous current policies represent a cruel deception.

## Notes

1. Relbec: *The Answer to your Questions about Industrialized Housing* (Hato Rey, Puerto Rico, n.d.).

2. Planning Office, Puerto Rican Urban Renewal and Housing Administration, *Industriaized Housing Systems for Puerto Rico* (San Juan, April 1971); Junta de Planificación, *Estudio para el Diseño de una Nueva Política de Vivienda para Puerto Rico* (San Juan, June 1971); Uriel Manheim, *Puerto Rico Builds: The Island's Housing Market in the 1970's* (San Juan: Housing Investment Corporation, 3d ed., 1972); Planning Office, P.R. Urban Renewal and Housing Administration, *Industrialized Housing: Multistorys—Puerto Rico* (San Juan, May 1972); W. Paul Strassmann, *Innovations in Building Methods and*

*Employment in Puerto Rico* (Geneva: International Labor Office, October 1973).

3. *Report of the Seminar on Prefabrication of Houses for Latin America,* United Nations, 1972 (ST/TAO/SER.C/141).

4. Santiago Luque Torres, "Clasificación de los Sistemas de Prefabricación en Colombia," *Memorias del Primer Seminario Latinoamericano sobre Prefabricación Aplicada a la Construcción de viviendas de Interés Social,* (Bogotá: Instituto Colombiano de Normas Técnicas, 1972), pp. 20-22.

5. Lucia Raffo de Mascaro, "La Prefabricación en la Producción de Viviendas Económicas—Análisis de la Situación Argentina: Propuesta de Política a Seguir," *Memorias* . . . p. 11.

6. Raffo de Mascaro, "La Prefabricación," pp. 37-38.

7. Otto Koenigsberger and Robert Lynn, *Roofs in the Warm and Humid Tropics* (London: Land Humphries for the Architectural Association, 1965), p. 12.

8. *Prefabrication* (November 1953).

9. *Colonial Geology and Mineral Resources,* vol. 4 (1954).

10. *Prefabrication* (March 1958).

11. R. E. Platts, *The Role of Plastics in House Structure* (Ottawa: Division of Building Research, 1964), p. 16.

12. *Plastics for Building in Developing Countries, Overseas Building Notes,* no. 134 (September 1970), pp. 6, 8.

13. Center for Development Technology, *Summary Report* (St. Louis: Washington University, 1972), pp. 10-12.

14. An analysis of other issues can be found in the author's *Building Technology and Employment in the Housing Sector of Developing Countries* (Geneva: International Labor Office, Technology and Employment Project, 1975).

15. *Towards Full Employment: A Program for Colombia* (Geneva: International Labor Office, 1970), pp. 171-72, 181-82.